To Phil Hamburger

One of class of '38s

very best

Smoky Ryan

JODI

Gengis Khan's DNA Progeny
Invades Washington

A Novel

by

I. M. "Smoky" Ryan

Published by

REMCO, Inc.
1339 Gunnell Ct.
McLean, VA 22102
(703) 790-9061

ISBN O-9623535-3-1

Printed in the United States of America
First Printing 1995

REMCO, INC.
1339 Gunnell Court
McLean, VA 22102.

ACKNOWLEDGMENTS

Several persons read and offered me encouragement. Colonel William Barney, (USAFRet.) was most helpful in meticulous reading, instructive ideas and offering suggestions on basic problems. Stan and Jane Doore for proofing; Jeanette Sterbutzel for editorial reviewing and editing; Joe Sterbutzel for providing computer expertise in formatting and preparation. My wife, Colleen, performed the final edit. My two boys, Kevin and Kirk, were both supportive. My heartfelt thanks to all.

ABOUT THE AUTHOR

I. M. "Smoky" Ryan attended the University of Nebraska where he was editor of the daily newspaper. He got his graduate degree from the Pulitzer Graduate School of Journalism just as World War II broke out. After serving eighteen months in the Aleutian Islands he returned to the states where for a short time he headed the Air Force Band on a tour recruiting WACs. All that was before D Day in Europe. Next with Henry Pringle he flew around Europe and the Middle East setting up Military Historical Commands at the behest of President Roosevelt. At Columbia he studied under Pulitzer Prize winners Pringle, who recruited him for the tour of top U.S. Commands, Douglas Southall Freeman and Walter B. Pitkin. He also studied under two Managing Editors of The New York Times, Bob Garst and Ted Bernstein. After a stint in New York with start-up airlines, Lt. Colonel Ryan was recalled to the Air Force during the Korean war. Later he owned night clubs. For 25 years he was a lobbyist and marketeer in Washington, DC He got interested in writing novels after listening to the trials and tribulations of those unfortunates forced to appear before Congressional inquisitions.

Publisher
REMCO, INC.
1339 Gunnell Court
McLean, VA 22102

JODI

Setting for JODI is the year 2000.

After an auspicious beginning, the Newt Gingrich programs got sandbagged by the return to congress of the ***'government owes me'*** crowd of political hacks. Much as discarded communists were restored to power in Soviet Republics by promising the good old days, Speaker Metzer surfaced as a throwback to Speaker Tip O'Neill.

Jodi's furious battle is an effort to carry forward the goals of the 1995 Congress, which suffered *defeat* after a bitter smear campaign waged by frightened bureaucrats in states dependent upon large federal payrolls.

Once again the electorate had proved that to sleep is to allow thieves to steal away the fruits of victory. In pioneer days Indians stole horses from sleeping sentinels. Bickering Republicans forgot to keep their sentinels awake. They allowed big spender demagogues with the loudest megaphones to recapture congress.

Jodi threw herself into a death struggle against the new congress and the old bureaucracy.

A traitor in the highest echelon of government; murder, foreign intrigue, Iranian assassins, the FBI, the CIA, beautiful women and powerful personalities drop all pretenses in a fight for supremacy.

Jodi, cowgirl from Nebraska, falls back on her Mongol DNA genes to ride into the battle. Calls for impeachment, an inept congress reminiscent of the discarded Continental Congress of 1776 and a declaration of National Emergency all serve to awaken Americans to the need of saving their collapsing democracy.

JODI is a novel about an extraordinary high-spirited woman against the background of a self-serving climate in our nation's Capitol.

ISBN 0-9623535-3-1

1. SPEAKER'S OFFICE

Her job in the Speaker's office only dampened the fires of revenge which burned deep in Jodi Zukov's belly. In the capitol, which sat like an isolated medieval castle on the highest point of land in Washington with the dirty moat of an unmanaged city surrounding its base, she still clung to memories of crimes and injustices committed in past years against her father and grandfather. But now, despite or because of such offenses, she had arrived.

Jodi Zukov loved her job. She had worked hard and pulled every string in the book to land the plum of working on Speaker "Fat Albert" Metzer's staff. Yet it would have been impossible without political intervention. Congressman Jim Pace, with Nebraska's small allotment already filled, traded a crucial vote for Jodi's job. It was part of the Speaker's modus operandi to use barter for control and power. Jodi and her father had been active vote getters in Congressman Pace's western constituency.

In politics high technology had yet to replace the primitive barter system so Congressman Pace rewarded Jodi for all her grass root efforts on his behalf the old-fashioned way. He handed

Speaker Metzer the critical vote needed to supply arms to Iran. Jodi learned early how the political system works.

At long last Jodi felt secure in a position to right some of the wrongs which had been inflicted on her father and grandfather by government bureaucrats. Her whole education had been tilted to this end. A law degree from the University had been implemented by tireless organizing in Nebraska politics. Her local congressman finally had been unable to ignore her insistent requests for a Washington job.

Jodi smiled smugly to herself as she laid copies of Monday morning's schedule on the Speaker's desk so he would have the information at his fingertips when he arrived late, which he always did after the weekend. Events were finally falling into place. Nothing seemed more assured than her meticulously crafted plans for the future.

She glanced at her wrist watch. Eight o'clock. The office was a tomb. Nobody worked late on Friday. Usually by middle afternoon everyone was gone. She sat back in the Speaker's chair and closed her eyes. The family struggle had been long but with her job in the United States Capitol her generation had finally made it.

Jodi Zukov's White Russian ancestors had fled and settled for a time in Mongolia; thence to Alaska, and finally to Nebraska where there was a large Russian colony in Lincoln. The vast Russian steppes held a common bond with the wide Nebraska plains. The Mongol blood carried in Jodi's veins had been picked up from Jodi's grandfather who married a girl from Karakorum. She inherited blue-green eyes from the White Russian side of the family and the black hair, soft copper-toned skin from her grandmother, a woman inordinately proud of being a direct descendant of Genghis Khan.

Though Genghis had many descendants, Jodi's father insisted his genes had duplicated themselves again in Jodi. The DNA had only lain dormant through the centuries.

Jodi's grandmother had amassed a library on Genghis Khan and discussions of his campaigns were a favorite topic of the family on winter nights.

"Knowledge was important to Genghis," she often heard her grandmother declare. "Just as it is today. Although he could barely read or write, his campaigns were brilliant. He sent out a network of spies and patrols and listened attentively to reports from caravan drivers, bribed merchants and scouts posing as traders. No bit of information was ignored."

The tradition of great horsemanship had never left the family. She and her father were regular participants in the rodeo circuit and state fairs held in Nebraska, Wyoming, Colorado and Montana.

One defeated opponent described Jodi bitterly.

"On horseback she's a damned savage. Like the *Devil's Horseman* riding in from the steppes. Her eyes get wild and she leans forward in the saddle hammering the horse until I think its lungs will burst. When her rope settles over a calf she yanks it on its back like hauling in an escaping mad dog. After tying all four feet into an immobile helpless clod she steps on the animal and waves her arms in conquest, face beaming and yelling like a victorious Mongol. In the rodeo ring she's unreal."

Jodi's new job had prevented this year's entry into the national rodeo in Cheyenne, Wyoming. A year earlier she had won the women's calf-roping contest.

As a teenager she herded cattle, rode fence, wore jeans and sat a horse like an Indian: no saddle, no stirrups, long legs dangling free and whooping it up at a full gallop. Finally her father admonished Jodi to use a saddle. He reminded her that stirrups permitted the control needed for weapons, roping or sudden maneuvering.

Jodi's biggest trouble had been with academics. One day in high school a teacher called her father for a conference.

"Jodi may not graduate unless she pays more attention in class and does her homework," Ms. Tafani said. "She doesn't concentrate on my lectures or cooperate in class assignments."

"She's a heavy reader," Boris Zukov protested. "Not much else to do on a ranch except work. We both constantly load up on reading materials from the town library—especially in winter months."

"Jodi's excellent when questioned on subjects," conceded Ms. Tafani. "She's bright but never quite focused."

"Not focused? What do you mean?"

"Not on homework. Not on class assignments. In a group task she's not interested."

"Such as?" questioned Jodi's father.

"Decorating the room for a stage play on an incident in history. Jodi sits in a corner and reads."

"Probably because she's sick of decorating. The two of us painted a huge cattle barn this summer. We decorated the entire area with a lighting system and installed new hardware and pulleys to operate a hay lift. She did more work moving stuff around with pulleys than your entire stage crew does in a year."

"In school Jodi should participate," Ms. Tafani said adamantly, feeling the interview was getting away from her.

"She's sent to school to learn."

"Cooperating with me and others is learning," Ms. Tafani said, her face reddening at Mr. Zukov's stout defense of Jodi's behavior.

Ms. Tafani was a new Iranian teacher not yet adjusted to the independent ways of American youth. She demanded strict obedience to the point of servitude which she had learned in Tehran. Her mullah father had been transferred to the U.S. only three years earlier.

Boris Zukov pulled at his chin thoughtfully before replying. It made no sense to further alienate Jodi's teacher.

"I'll speak to Jodi," he said. "I never realized she had a problem with cooperation."

At home Boris Zukov questioned his daughter about her attitude in school.

"Simple," said Jodi. "Ms. Tafani dislikes me."

"Why?"

"Doesn't approve my sense of humor. Thinks I made her a laughing stock in front of the class. Sent me to the principal's office."

"You never told me?"

"Wasn't important."

"What happened?"

Jodi gazed lovingly at the concerned look on her father's face, its lined leathery skin abused by the sun and wind. She laughed.

"Ms. Tafani has a habit of droning on and on about a subject...keeps repeating herself. I have trouble not falling asleep." Jodi paused, bringing back memories.

"Yes. I'm listening," prompted her father.

"Ms. Tafani keeps a tall thin vase on her desk with a single flower, which she meticulously replaces every morning," Jodi said, a slight smile twitching the corners of her mouth. "One day someone knocked it over and spilled water during the lunch hour."

"She had a point in getting miffed."

"Okay," said Jodi, her dimples deepening and her blue-green eyes lighting up. "But she kept demanding to know the culprit, who obviously was afraid to speak up.

"'Come now, whoever knocked over my vase knows. They set it back upright,' Ms. Tafani kept harping, stubbornly going over the incident four or five times. Finally the repetition was getting on my nerves.

"Maybe the water evaporated into a cloud and rained, I foolishly announced to the class, which responded with whoops of delight. The next minute I was headed for the principal's office."

"Not a brilliant act of diplomacy," responded her father with a tight grin. That was all he ever said about the matter.

What Jodi didn't tell her father was the aftermath of his visit. Ms. Tafani became enraged with her inability to control Jodi. Finally, near the end of the semester, she notified Jodi her insubordination forced a failing grade.

"Why?" demanded a furious Jodi. "My lowest test grade is a B minus?"

"Deportment unbecoming a young lady," said Ms. Tafani fiercely. "You tried to mock me all year long."

"Not true," protested Jodi. "It's your vivid imagination because I didn't care for your Muslim lectures."

In the two weeks remaining all arguments proved futile. Jodi sat in class without saying a word because she intended to appeal the action and she didn't wish class absence used against her. In the middle of the final week a frightening event changed everything.

Ms. Tafani sat down at her desk after giving her usual ten-minute dialogue on Middle East history. She pulled opened her top desk drawer and screamed in terror.

A huge barn rat jumped on her bosom, climbed on her shoulder and clawed its way into her fastidiously coiffured hair. For several minutes it furiously fought entanglement. Initially, it was hard to tell who was in greater terror, Ms. Tafani or the rat. After destroying the hairdo the panicked rat managed to extricate itself and bounded to the floor and out the partially open door.

Ms. Tafani continued to scream hysterically until she finally fainted, falling face down on the desk. An ambulance rushed her to the hospital where a doctor kept her on sedation for three days. She never returned to school the remainder of the year. The principal finally announced she had left the states and moved back to Tehran where she had relatives.

No one established who brought the rat to school. For lack of proof the matter was dropped. After reviewing Jodi's exam papers, the principal revised Jodi's grade to ***B-minus.***

One other incident occurred while Jodi was a sophomore in high school. A senior jock on the football team named Roy Wolfe knocked Jodi down while running in the hall. He stopped to sneer, looking down at Jodi.

"Stupid bitch, get out of my way."

Proud owner of a robin-egg blue convertible he came out after practice one evening to find all four tires flat and the air valves lying neatly on white napkins beside the tires. Although the tires weren't damaged it took a good hour of hard work with a hand pump to inflate them. On the steering wheel he found a note.

"You should be thankful a *lady* let the air out," the note said. "A *stupid bitch* would have slashed tires with a knife, used an ice pick or a hammer and nails. Then you'd had to buy new tires."

Wolfe stood scratching his head until he remembered. The tomboy he'd accidentally decked. The one who rode a horse to school each day.

The following weeks he could barely keep his eyes off Jodi. What a reclusive odd-ball. For the first time he noted her classic beauty, camouflaged in boots, jeans and cowboy shirts. Slowly he decided she was the most delicious female he'd ever known. She rarely wore a dress. Occasionally a leather skirt. But the cold-steel eyes somehow didn't belong with that face. An uneasy Roy Wolfe couldn't determine why. He learned to his own chagrin what other kids instinctively knew—Jodi was not one to fool with.

She rode across open fields and jumped fences to get to school each day. Saved three miles from the five it took by road.

Late in the year Roy Wolfe confronted Jodi.

"Understand you're a lady," he said with a wide grin. "May I take a lady to the senior prom?"

Startled, a surprised Jodi surveyed Roy carefully before answering.

"As a lady to a gentleman, thanks for the invitation. But no. My father thinks a sophomore dating in high school is too young. I agree. I have too much to do on the ranch."

Roy's disappointment was obvious.

"Oh."

Jodi said. "Perhaps we'll meet in college? If so, as a gentleman you might try again after I learn to dance." She walked away.

Roy Wolfe looked after her as she disappeared into a classroom.

Weird," he muttered. "What a weird one."

*   *   *

Jodi was startled out of her reverie by a booming voice.

"Fall asleep? That how you work after everyone's gone home?"

Ali 'Knuckles' Rahman, former Chicago Bears defensive end and now an administrative aide to the Speaker, filled the doorway. Blocking the opening, 'Knuckles' looked about as inviting as an alligator in a children's wading pool. Jodi's skin crawled. He filled the room with the dark odor of his famed cruelty. Heavy fines and threatened expulsion had not lessened his terror to opposing quarterbacks, for before the rules changed he usually managed to accidentally break or dislocate something with each sack.

Jodi, a bit disoriented, sprang to her feet.

"Just finished the Speaker's schedule for Monday."

Rahman grinned understandingly.

"Don't worry. I won't tell the boss. Take a catnap myself whenever I can."

Jodi moved around the desk to the sofa where she had left her purse.

"I've finished." She looked up at Rahman. He hadn't moved. "Anyhow, what brings you into the office? You're rarely here during working hours so you must have forgotten a free ticket."

Jodi resented the fact Ali Rahman was the only one in the Speaker's office who worked when he pleased. Ali's father was a powerful Ayatollah in the heavy Muslim population in the Speaker's district in New Jersey. Muslims supported and advanced Speaker Metzer money for his campaigns. In return he was a strong champion of dealing more favorably with Iran. One day she inadvertently saw the Speaker accept a fat envelope just prior to a critical arms vote. She heard the senior Ayatollah Rahman's instructions, "spread this where necessary."

Ali silently watched Jodi gather her handbag. He knew a great deal about Jodi. Most of the office did. Background files were no secret to an office staff.

In University Jodi had been a loner. Her grandfather had accumulated a large spread in Nebraska when land was cheap. It was rumored he had bought the land with jewels and heirlooms spirited out of Russia. He had been a horse trader as well as a rancher, traveling miles to attend stock auctions. Jodi's father had enlarged the operation by establishing feed stores throughout the midwest.

During the cold war against the Soviet Union, Jodi's grandfather told stories of persecution by zealous U.S. bureaucrats.

Later her father faced horror years of bitter lawsuits and litigation initiated by the government. The Environmental Protection Agency confiscated land and levied big fines under a wetlands verdict involving a minuscule ten acres of swamp which had been filled in and converted into grazing land for horses and cattle. That the swamp had no flowing water and was a breeding ground for mosquitoes and other pests harmful to livestock and humans made no difference to EPA. Zealots even tried to send a man to prison for killing a rat. In Washington the bureaucracy lives and thrives. The Czar's most abusive tax collectors never equaled EPA's moronic efforts.

A long court battle and her grandfather's death ensued before Jodi's father finally got part of the land back. Meanwhile the protracted fight and legal fees left her father broke. Private lawyers demand pay up front. Government lawyers are paid by the taxpayers to litigate against taxpayers.

Jodi's father was barely on his feet again when bureaucrats dumped crushing new financial regulations on him. With hard work and long hours he had built an organization of 156 employees in his feed stores. New laws said he must comply with minority regulations. In fourteen years of traveling throughout eight midwestern states he had never known a black farmer, yet he was heavily fined for not hiring one.

To Jodi's young mind plundering bureaucrats had become the essence of democracy.

Jodi grew to hate bureaucracy. Her original reason for going to law school was to learn how to fight government lawyers and help her father. Oppressive laws twice had nearly destroyed her grandfather and father.

Ali Rahman knew much of Jodi's background, but not the underlying details. He knew she had few friends. Without question she was vulnerable.

"No use carrying a chip on your shoulder," Rahman said. "That'll get you nowhere in this environment."

Jodi faced him, lips quivering.

"I give full measure to my job. I've no apologies on that score."

"You'll advance faster if you loosen up...socialize more," Rahman said, closing the door behind him and moving towards her.

"I do my work." Jodi edged around the coffee table in front of the sofa.

"Slaves do their work. But it's no fun." Rahman kept himself between Jodi and the door.

Jodi became nervous, like a colt with a bit slipped into its mouth. Although there was no reason to fear a fellow employee, Ali Rahman was huge and he seemed to be subtly impeding her path to the doorway. In pro football he had delighted in the nickname *Mean Ali Rahman.* It had enhanced his macho image to accidentally maim an opponent.

"I sought this job. Slaves don't choose their work. Anyhow, I've got to go." Jodi smiled disarmingly. "Hope you find your ticket."

"Thought we'd have a little chat. Allay some of the animosity you hold against me. Reach an understanding."

"I hold no animosity. I just wish you'd leave me alone."

Rahman didn't budge. He stared at Jodi. Her presence in the office was tantalizing. Nice brown legs and shining coal black iridescent hair falling in soft waves on shoulders broadened by ranch work.

Jodi wore a short purple skirt, her father's favorite color since it was dominate in his ancestor's Coat-of-Arms, a soft white silk jacket open at the throat and a braided leather belt cinched around the wasp waist. A purple head band completed the ensemble.

Yet it was not the structured beauty that made Rahman's libido rise like a summer heat wave. Nor was it the blue-green eyes set in contrast to her warm skin. It was her mouth. It was the most sensuous mouth Rahman had ever seen. The lips were full, curved, plush and exuded promises of fleshy delight.

"First a kiss," said Rahman, stepping forward. "Help you lighten up and join the social world."

"Don't be stupid," said Jodie, attempting to swing around his huge bulk. "You know the rules. Those in the same office don't fraternize."

"Evidently you've never heard of General Patton's rules. Copulation is not fraternization."

"What the hell you talking about?" Jodi made a determined plunge for the exit.

Rahman grabbed her wrist. Jodi looked down at his hand. It circled her arm like a python. The blood drained from her face. She suddenly felt fear. The only blemish on her skin was a long thin scar which ran from below her right ear lobe up behind the length of her ear into her hair line. The ear had been nearly torn away from her head during a calf-roping competition in Cheyenne. After doctors sewed the ear back on she returned to win a quarter-horse sprint the next day, her head swathed in bandages and whooping it up across the finish line. Now that scar turned chalk-white and the area began to burn.

"I'm talking about getting off your high horse," said Rahman, planting his right hand behind her head and tilting her face upward. As he bent over to force a kiss Jodi violently swung her head, smearing lipstick across his face.

"Want to play rough, huh. I've played against tougher Mongols than you all my life."

Jodi's resistance excited Rahman into a heightened rage. He picked her up like a limp dishrag and threw her on the sofa.

"Time for you to learn what life's all about."

He fell heavily across Jodi. As she started to scream Rahman placed his immense hand across her mouth.

"Holler and I'll break your jaw. Besides, who's to hear. Place is deserted and I've locked the door."

Rahman's huge body was crushing the wind out of her. She could hardly breath.

"Please. You're hurting me. I'll be good," Jodi gasped, helplessly realizing that further struggle would only bring bodily injury.

Rahman's face was livid with desire. He started tugging at the top buttons on her blouse.

"Once we've made love, you'll change your tune. You'll probably even want more. They usually do."

"Okay...Okay," Jodi blurted. "Only let me catch my breath. You'll enjoy it more if I cooperate."

"You mean you'll cooperate?" Rahman lifted himself above her but maintained his hands on both sides of her body.

"Since I've got no choice, why not? Just be gentle."

Rahman studied her suspiciously for a moment, then rolled over to the side, sliding off his trousers with one hand.

"That's more like it," he said with a smug smile.

Jodi started unbuttoning her blouse, bringing both hands up to her shoulders. Satisfied, Rahman sat up to take off his shirt and slip out of his shorts.

Ali felt no remorse in violating Jodi. Raping an infidel woman was no crime, according to the legacy of Khomeini.

With Rahman's attention momentarily diverted, Jodi reached above her head to the end table adjacent the sofa. Quickly she grasped Speaker Metzer's favorite ash tray—a heavy solid glass base with an exquisite hand-blown peacock on one side. The Speaker had received it as a gift from the French on one of his overseas jaunts.

When Rahman turned to see why Jodi was squirming away, she swung the ashtray with all the force she could muster against the side of his head. The peacock splintered into a thousand pieces as the heavy base crashed into Rahman's temple. Blood spurted over her hand as Rahman collapsed to the floor, suffering a temporary concussion.

Jodi immediately sprang to her feet and fled to the door, unlocking it before turning to view the moaning Ali, naked except for his shoes and socks.

"You bastard," she said between gritted teeth. "I hope you bleed to death. I'm buying a gun. You ever try to rape me again I'll fill you with so many holes you'll think you're some form of Swiss cheese."

"You bitch," Ali muttered in agony, placing his hand against the side of his head and staring at the blood flowing over his fingers. "You Mongol bitch. I'll get you for this if it takes a thousand years."

Jodi slammed the door behind her and walked down the long corridors with misty unseeing eyes. The click of her heels against the sound of marble was the only noise in the building.

2. CONFRONTATION

Late Monday morning Speaker Metzer returned from committee meeting and called Jodi into his office.

"I've had disturbing reports about you," he said without any preliminaries.

"Yes, sir," was all Jodi could say. She stood nervously across the desk facing the Speaker.

"Ali Rahman got out of the hospital yesterday with 16 stitches in his head. You broke my French ashtray against his temple." The Speaker looked at Jodi in disgust. "That sharp pointed peacock could have penetrated an artery, then you'd be facing a murder charge."

"He tried to rape me."

A surprised look crossed the Speaker's face. He quickly recovered.

"Ali tells a different story."

"I'd expect he would."

Jodi remained silent as the Speaker looked away from Jodi and stared at the ceiling, collecting his thoughts. He spoke in a strained voice.

"Ali says he caught you going through my desk drawers and when he tried to interfere you hit him with the ash tray."

"Ali's a liar."

Speaker Metzer persisted, rumpling his white shaggy hair in a fit of exasperation.

"He told me you were seated behind my desk with the drawer open. What were you looking for? Confidential correspondence?"

"I wasn't looking for anything. I'd just placed today's schedule on your desk."

"There was no schedule on my desk this morning," said Metzer, looking down at his clean desk.

"Then Ali's destroyed it."

"That's absurd. Why?"

Jodi's temper rose. She could feel the flush rising up her neck into her ears. Obviously the Speaker didn't intend to believe her.

"How do I know, Mr. Speaker? Probably to bolster his yarn that my only reason for being there was to rummage your desk."

The Speaker leaned back in his chair and stared long and hard at Jodi. His heavy jowls shook with anger.

"You know you're creating an ugly story? If this thing breaks out of the Speaker's office, the media will have a field day."

"I've no intention of telling the media anything. Just fire Ali and the incident will be forgotten."

"I can't do that. It's his word against yours."

"We can't work together in the same office."

"I realize that," said the Speaker, his eyes turning hard. "Ali's been with me five years. You less than six months."

Jodi suddenly grasped the Speaker was playing a cat-and-mouse game, trying to twist the facts. He had no intention of giving her a fair shake.

"You plan to fire me?"

"What other option do you leave me?"

"Face the truth, Mr. Speaker. You've told the nation a thousand times your motto is to seek the truth."

Speaker Metzer's face reddened. "I don't think it's your place, young lady, to lecture me on the truth."

"If you fire me," Jodi said, her chin determined but quivering, "all bets are off. I don't intend to go quietly. I'll demand lie detector tests for both Ali and me."

The Speaker's face blanched. For the first time he realized the young snip before him was going to spit and claw even if it meant destroying his office. His demeanor changed.

"You'd put on a phony crying jag for the media? That's old hat. The public now knows women lie. Susan Smith pleading tearfully for a black carjacker to spare her children? Anita Hill rehashing lies against Supreme Court Justice Clarence Thomas? Another woman beat her daughter to death, buried her and then bled tears to the public?"

Speaker Metzer took a deep breath and exploded.

"Lies can't hide behind skirts anymore."

"Voters also know politicians lie," said Jodi. "We'll have a front page bonanza. F*emale staffer charges rape in Speaker's Office*. I'll take my chances with the truth."

Metzer rested his head in his hands with both elbows on the desk. His flaccid features reflected his agony.

"There's got to be compromise," he said at last. "We can't allow a scandal to grind the work of congress to a dead halt for the next three months."

Jodi remained silent, gazing out the windows behind the Speaker. She'd spoken her piece.

Speaker Metzer gritted his teeth and turned to the attack one final time. Perhaps Jodi would find the atmosphere unpalatable and resign of her own accord.

"You know you were practically forced down my throat to begin with?"

"I know you hired me as a favor to Congressman Pace from Nebraska. He said you owed him a chit."

"I owed him nothing," the Speaker snapped. "I was just doing him a favor."

"Whatever the reason, I was glad to get the job. And I've done my very best to handle it well."

The Speaker's many years of negotiation had taught him the first step was to undermine his opponent's strength.

"I've studied your resume, Miss Zukov. The FBI filled me in with details..."

"Obviously there was nothing wrong or I wouldn't have been hired," Jodi interrupted for the first time. She intended to stand her ground. Clasping her hands behind her back she dug fingernails into the palms of her hands until they were numb with pain. They diverted pressure from the Speaker's grilling.

"I've been here nearly three decades, young lady," the Speaker said in steely tones. "Before you jump on your white horse let me tell you something about yourself."

"Yessir," said Jodi, shifting her feet to ease the tension.

"You're an ambitious tyke with a craw in your throat. Your belly's on fire from imagined past wrongs. Your parents and grandparents waged war against this government despite the fact your ancestors fled Russia to seek refuge here. What are you? Ingrates?"

"We're not serfs," Jodi retorted. "Most Americans fled from other countries."

"I've seen your type come and go in this city. Fanatically ambitious. Crusader. Intolerant of bureaucracy. With a chip on your shoulder."

"If George Washington, Thomas Jefferson and James Madison hadn't had chips on their shoulders, you wouldn't be Speaker today." Jodi fought to control her anger. "I'm glad you cast my family as ingrates because they fought stupid regulations. That puts them in a class with our founding fathers."

A flash of admiration sparked the Speaker's eyes. The brat had spunk. The moment passed.

"Could be your family's been wronged. The EPA and DEA were like putting elephants to roam in a peanut field..."

"Where they still trample at will," exclaimed Jodi.

Speaker Metzer grimaced at the interruption.

"That doesn't change things," he said. "Regulations are the law. You can't personally decide which laws to accept or reject."

"I intend to fight unjust laws."

“Well, I certainly won’t allow you to use the Speaker’s office as a bully pulpit for personal grievances.”

“I’ve never done so, sir. You’re twisting this interview.”

“You’re basically an arrogant maverick,” sputtered the Speaker.

“That may be true,” said a tight-lipped Jodi, “but it has nothing to do with attempted rape in the office of the Speaker of the United States.”

“Congress faces another tough election year. This is no time for a wild-ass story like yours to hit the street.”

“Dad was right about congress,” said a defiant Jodi, a tear trickling down the side of her cheek.

“Yeah. An immigrant rancher from Russia knows all the answers.”

“You in congress just don’t get the message.”

“A message from the last Czar? Or Stalin?”

“When a pilot gets drunk and kills 200 people the public sues the airline,” said Jodi, her back stiffening. “Furthermore, they refuse to fly the airline until in some cases it causes bankruptcy. When a postman robs or dumps mail in a garbage can the public hue and cry is over the lousy postal service. Cartoons depict an organization out of control. Same with an Exxon oil spill. Not one person in a thousand can name the pilot, postman or ship captain.”

The Speaker interrupted impatiently.

“What the hell you driving at?”

“You in congress still can’t grasp what’s happened?”

“You don’t make sense.”

“Like dad said,” Jodi exclaimed between gritted teeth, “you encouraged and allowed government agencies and bureaucrats to brow-beat, malign and abuse citizens. Voters hold congress responsible for the impenetrable wall of an insensitive, arrogant and stupid bureaucracy. So citizens rightly become enraged at the boss.”

“Nonsense.”

Jodi decided that since she was certain to be fired she might as well let it all hang out.

"Until congress quits coddling such conduct the voter will continue to rage against the instutition which creates such wayward zealots."

"You're avoiding you own problem." Speaker Metzer's eyes flashed cold rage.

Jodi plunged ahead, determined to get her father's bitter experiences off her chest.

"Long as congress creates Spanish Inquisitors you'll have to bear the onus of your own misbehavior. Congress creates bureaucratic monsters, then lets them roam the country like wild dogs without leashes. Your zealous bureaucrats stick their malevolent noses into the lives of every citizen. The people hate them."

Jodi looked into the Speaker's angry face.

"Congress still hasn't got it," she said. "But the voter's got it up to his eyeballs. He resents a body which lets bureaucrats urinate wherever they choose. Congress best rein in their bureaucratic rogues if they expect to achieve any bond with the people."

Speaker Metzer became distraught, almost to the point of losing control. His flabby bulk shook with frustration. For a full minute Jodi thought the Speaker would break into tears.

"Unless you desire to destroy this office, I propose a solution," he said grimly, eyeing Jodi with the hate of a conquered gladiator.

"I've no desire to damage your office, sir," Jodie said sincerely. "You did give me a job. But I'll not become a scapegoat."

"If you quietly resign I'll see that you get one full year's severance pay? That should get you back to Nebraska with money in your pocket."

Noting the look of hesitation on Jodi's face, Metzer added.

"No one need know anything. Obviously you can no longer work with Ali. Yet firing him would have repercussions from here to eternity. The media and a dozen select committees would jump into the fray. You'd be set adrift without a dime and lawyers in this town demand up-front money—lots of it."

The Speaker fidgeted with the white carnation on his lapel. Conceived to identify congressmen at fund raisers, Speaker Metzer wore the white flower in his buttonhole constantly. It announced his special status and whenever nervous or at a loss for words he habitually lifted it to his nose to smell the fresh fragrance. Twice daily he changed the flower to preserve the freshness.

Jodi studied the Speaker's face. His fidelity could never be questioned. He always considered his own interests first. One always knew where he stood. He had one foot planted solidly on each side of any issue until he saw which way the wind was blowing. At the moment it certainly wasn't blowing in Jodi's favor.

Jodi knew the Speaker spoke the truth. She would never again have a moment's peace. Worse, she would be plastered by some horrible label dreamed up by the media. Any possibility of a career would be cut off at the roots. Past experiences of Jodi's family against the labyrinth of government proved a single individual had the effective clout of a gnat beating against a plate glass window. Individuals were invisible. Only causes, organized voting groups or raw money garnered attention in Washington. She saw no recourse but to accept the Godfather offer.

The harassment issue had started to boomerang. It yo-yoed with the times. After the feminist movement supported Anita Hill's lies, the public got its fill of trumped up accusations. The Speaker was right. People no longer believed every tearful victim seeking publicity. Futhermore, the publicity rarely helped the victim.

Jodi said. "If you're going to wreck my livelihood, I'm going to return the favor. Just because you're Speaker this is one little nobody you're not going to tramp on."

"Giving a year's pay to a tramp hardly seems I'm tramping on anybody."

A warning flashed in the back of Jodi's mind. Past Washington scandals on payroll shenanigans ran up a red flag. Prosecutors hanged victims along with the scoundrels.

"Where will you get the money?"

"None of your damn business." The Speaker ran a frustrated hand through his hair. On second thought, trying to end the uncomfortable session, he added, fuming. “From my staff allotment. I’ll just have to hire one less person in my office for a year. I’ll take care of it. That’s not your concern.”

“I’m afraid it is,” said Jodi.

Speaker Metzer half rose from his chair, looking over his desk to find something to hurl at the stubborn wench who was destroying his peace of mind. He spoke in a choked voice.

“Like hell it is. Take my offer or clear out.”

“Accepting a year’s salary on your payroll for doing nothing is against the law.”

Jodi’s statement brought the Speaker up short. She was right. He knew better. His rage had warped his judgment.

“Can’t take it out of campaign contributions,” he said, settling back into his chair with a frown.

“I know that too,” said Jodi.

“Then where the hell...?”

Jodi supplied the answer.

“From your own pocket. Where it hurts. Where you’ll experience some of the pain you cause others.”

Speaker Metzer stared at Jodi in disbelief.

“You’d leave me with nothing to live on?”

“About one-fifth of your salary, not including perk and pork. You rarely reach into your deep pockets anyhow. I think you can manage.” For the first time Jodi felt she was meeting the Speaker on equal ground. Her spirits rose.

“Impossible,” Metzer blurted, feeling like a cornered rat before this obstinate nemesis.

Jodi said. “Your one-fifth pay still amounts to more than many Americans make per year. Will give you a better insight into what you leave taxpayers to live on.”

Speaker Metzer glared at Jodi for a full minute.

“Okay, okay,” he said. “Worth a personal check to get you the hell out of this office. Just never cross my sight again or...” He left the threat hanging in the air.

Jodi heard a snicker. She looked quickly and saw a crack in the door to the AA’s office. Ali had been gleefully listening. She realized the Speaker had intended to stage an act of dismissal for Ali to relate the event to his father. Probably for a bigger contribution. Unfortunately, the episode got out of control, not following the programmed script.

"Close the door, Ali," Jodi said disdainfully. She turned to the Speaker.

"You are a miserable man, 'Fat Albert'." Jodi deliberately used the expression which infuriated him. She headed for the front exit, stopping with her hand on the knob.

"I'll call at the House Post Office for my check Friday. It'd better be there in total or I'll have an earth-shaking press conference Monday." With those parting words Jodi flung open the door and was gone.

Speaker Metzer heard the sharp staccato sound of leather heels fade down the hallway. He wiped the perspiration from his forehead with the back of his hand.

"A bitch, a viper," he muttered as his AA emerged from the back room with his head swathed in bandages.

"A goddamned Mongol barbarian," said Ali, adding with a sly smile. "I'll get dad to cover your costs. I need the diversion of another trip to Switzerland."

"Thank God I've seen the last of her," said the Speaker.

Ten days later Jodie Zukov deposited the Speaker's check and moved into an abandoned second-floor office three blocks from the Supreme Court building.

She knew the Speaker would be furious when he discovered she had stayed in town. But she was determined not to be driven out of Washington. The humiliation of returning home would be unbearable. She'd heard there were over 40,000 paid lobbyists in the city. Surely one more would not overload the boat. New clients clamored for good lobbyists every day. Jodi had seen them crop up regularly in the Speaker's office.

Jodi crossed her fingers and gave a silent prayer as a painter lettered her office door—JODIE ZUKOV, ATTORNEY AT LAW.

3. ON HER OWN

Jodi spent the first week in her new office attending second-hand furniture marts. She succeeded in securing an old congressional desk and filing cabinet plus a telephone with two lines. She hoped to obtain a few small lobbying contracts for starters. She knew enough about the business to know industry didn't hire neophytes without connections. Like the chicken-and-egg. It didn't matter which came first. She needed both.

Jodi determined her best initial opportunities were women in congress. Perhaps they would be sympathetic and helpful. Women in the senate offered the best hope since they were fewer in number and she could concentrate her efforts.

Jodi still burned over the treatment she had received from the Speaker. It had been unfair. She had been placed between a rock and a hard place. The Speaker wanted her gone and the weapons to fight back were repugnant to Jodi. They involved scandal which could do her fortunes nothing but harm.

Jodi swore silently as she sat flipping the pages of the Federal Directory. A department of government had stabbed each generation of Zukovs. Her grandfather during the cold war, her father by EPA and now she had been discarded like a smelly dishrag by the most powerful man in congress. Her mind seethed with anger at the injustice.

Three months of pounding the pavement and making calls on industry had netted three clients, each lasting less than a week. She no sooner prepared to go to work than the client suddenly changed his mind. The excuses were weak but consistent.

"A review of our budget requirements prevents us from taking on new employees," said one. "I'm sorry but I should have checked financial before making any more hires."

"Head office in Chicago turned me down on hiring more personnel," said another, although Jodi personally knew two young men taken on after her rejection.

One client was particularly blunt.

"We checked with your former employer and Speaker Metzer's Administrative Assistant said you were fired for rummaging through off-limit classified papers. This company can't afford the Speaker's animosity."

Her eyes came across a picture of May Trudeau, Senator from Nevada. A media favorite. Glamour gal who regularly fronted such publications as Vogue and Cosmopolitan. A maverick who wore short skirts and high heels to the consternation of many of her fellow senators. Since the Speaker had accused Jodi of being a maverick, perhaps Senator Trudeau was a kindred soul.

Another three weeks passed before Jodi finally got in to see Senator Trudeau. Meanwhile, she left calling cards all over town. She got 12 immediate fund-raising responses from congressmen and placed on innumerable mailing lists. Each of the responses was an invitation. The cheapest was for a reception costing $200. The most expensive was an opportunity for monthly luncheons, ala the former Senator Bentson, with a yearly price tag of $10,000. None offered a job.

Senator Trudeau was gracious.

"Sorry I couldn't meet with you sooner," she said as she shook hands. "Was in Nevada most of the time tending home base." She waved Jodi to a seat on the sofa. "Care for a cup of tea?" She nodded to her assistant as Jodi accepted. "I gave up on coffee when I started climbing the walls."

Jodi was astonished at Senator Trudeau's beauty at close range. May Trudeau had a tight, compact curvaceous body with a healthy bust line. She carried her dark brown hair swept back and deftly tied with a red-and-white bow. A snug fitting imitation tiger belt cinched her trim waist.

In social circles May would hardly pass as a U.S. Senator. Proudly feminine, she defiantly wore spiked heels and skirts alarmingly high for a politician, refusing to yield to any artificial dress code. Despite her flamboyance, she was the darling of environmentalists. Her husband, a part-time rancher, chain store and land owner in Nevada, had first tested her mettle when trying to buy restricted land for a major-sized mall. He lost his mall but ended marrying May. Environmentalists energetically and enthusiastically supported her causes.

"What can I do for you, Miss Zukov?" said Senator Trudeau, accepting her cup of tea and signaling her receptionist to close the door.

Jodi shifted uncomfortably. She'd prepared a standard introduction but immediately sensed it wouldn't work with Senator Trudeau. She decided to shoot with the truth.

"I've started my own lobbying business without any clients," Jodi said in a rush. "I thought a major step ought to include meeting the people who I intend to lobby."

Senator Trudeau's look of astonishment changed to deep laughter. She dabbed her eyes with a Kleenex.

"That's the most honest approach I've heard since I've been in the Senate. Either you are quite clever or terribly naive. Senators have nothing to do with industry's selection of lobbyists."

"I'm not very clever or I'd not be out of work," Jodie said with a forced smile. "I worked on Speaker Metzer's staff for six months until I opened my own office."

"Yes. We've all heard of the Nebraska beauty who worked for the Speaker and suddenly quit. So you've decided to go out on your own, Miss Zukov?"

"Jodi, please?"

"Jodi?" Senator Trudeau pursed her lips. "A nice name to jump into a cage with the get-rich-quick wolves."

"I didn't exactly jump, Senator. Anymore than the Christians jumped into the Roman arena."

"Oh!" Senator Trudeau studied Jodi quietly before speaking. "I've no wish to know the details, but if it's any consolation I'm not a particular friend of the Speaker either. We've had some pretty brutal run-ins."

"I can only say our parting was mutually acceptable."

May nodded with a smile. "I understand. But helping you find a client, that's out of my bailiwick. Industry picks its own."

"I know that," said Jodi, getting up to leave. "Secretly, I just wanted to meet you anyhow. You're something of a legend to young women."

"Sorry to hear that."

"Sorry?" Jodi's puzzled look caused a smile to cross Senator Trudeau's face.

"Yes. Sorry," laughed Senator Trudeau. "Legends are for the dead. I still fancy myself as a young woman."

Jodi blushed her embarrassment. "I meant..." she stammered, groping for words.

"You meant it as a compliment," Senator Trudeau's dancing eyes revealed her merriment. She walked Jodi to the door.

"Thanks for your time, Senator," said Jodi, turning to shake hands."

"My door's always open to you, Jodi. With your beauty and education—I did read the resume you sent—I'm sure you'll find something better than lobbying. It's a rather unappetizing brown-nosing profession to which I somehow doubt you're suited. I know casinos in Nevada who would lap up a striking legal beauty such as yourself."

Jodi was to remember those words.

At home that evening Jodi received a call. Ali Rahman was on the phone.

"Landed any rich clients yet?" His voice was triumphant.

Jodi knew she should hang up, but she couldn't resist.

"Gutter work suits you, Ali. You've found your calling. I recognized your hand in getting me fired after I'd found a client."

"You were supposed to go back to Nebraska."

"I made no such promise. I intend to speak to your boss about your conduct. I'm sure the Speaker wouldn't sanction such sleaze coming out of his office."

Ali Rahman's loud laugh forced Jodi momentarily to hold the receiver away from her ear.

"You stupid broad," Ali said. "You just don't get it. The Speaker wants you run out of town. He's the one who assigned me the job of bird-dogging you into any client's front door."

"You're both bastards."

"You're the one headed for the garbage dump. I track you down but the boss calls your new clients personally. Says you're persona non-gratis on the hill."

"I'd no idea the Speaker could stoop to such sleaze."

"Baby, face it. You're a pigeon with two broken wings. You'll never fly in this town."

"That why you called, Ali? Just to gloat?"

"A damn fine reason in my book," Rahman said. "But I really called to remind you the severance pay was to leave town. Go back to the farm shoveling shit. That's where you belong, you arrogant bitch," laughed Ali as Jodi slammed down the receiver.

Despite her brave words, Jodi felt a crushing helplessness. Against the Speaker's malevolence, the chance of working in Washington was hopeless. She walked over and threw herself across the bed. For the first time since her mother's death tears streamed down Jodi's cheeks. The thought of returning home as a failure doubled the agony.

Good God, she anguished. What sins have I committed. Or is the Zukov family as jinxed by U.S. government as it was by the Soviets?

4. WASHINGTON, D.C.

Spring dragged by without Jodi landing a single client. She was using money at twice the rate she had anticipated. Another two months and she'd be on her way back to Nebraska like the whipped dog the Speaker gleefully expected. She would rather jump off the Washington Monument than suffer that fate. The phone rang.

"Name's Tony Pellini," said the voice, sounding like it came from the bottom of a barrel. "This Miss Zukov?"

"It is."

"Senator May Trudeau suggested I call."

"She say why?" The caller's voice disturbed Jodi. It sounded mean and harsh.

"I'm from Nevada. Need a small office. Senator Trudeau said you had space."

How on earth would Senator Trudeau know she had extra room? Besides a small empty space for a receptionist—which she couldn't afford—Jodi had a side room containing a two-drawer legal filing cabinet. Jodi guessed the senator had her spies and was genuinely concerned about how she was doing. Twice the senator had called but each time Jodi had put on a brave false front. Evidently Senator Trudeau hadn't been fooled.

"Depends," said Jodi cautiously. "When do you want to come over?"

"Right now, if convenient. I'm in a hurry to settle."

"Okay." Jodi glanced at her wrist watch. "It's 10:30. I'll be here another hour."

"Thanks. Be there in twenty minutes." He hung up.

Jodi was shocked at Tony Pellini's appearance. He had the voice of a bull but the wiry slender structure of a rodeo rider. She noticed his tall figure widened into muscular shoulders and arms, accentuated by a short-sleeved wild-colored silk shirt.

"I work for Guy Ericson," he said. "Guy's not happy with way things going in Washington."

Jodi's raised eyebrows and blank stare annoyed Tony.

"Never heard of Guy Ericson?" His shocked expression conveyed the thought Jodi probably had never heard of Ronald Reagan or Howard Hughes either.

"Afraid not. Have I missed something?"

"Guy owns the Palace Casino plus others in Las Vegas. The Palace is the biggest in the world. He owns a west coast pro football team and the REPUBLIC studio in Hollywood." He cast a jaundiced eye at Jodi. "He also owns 51 percent of the Hollywood Race Track, and you never heard of him?"

"Sorry."

"Where you from? Mars?"

"Nebraska."

Tony Pellini shrugged his shoulders.

"Explains everything. Grow anything besides weeds in Nebraska?"

"Pretty damn fine football teams, something I'm sure your boss knows about."

Tony bit his lip in thought. He smiled for the first time.

"Yeah. Boss got some good players from Nebraska." He walked around the corner into the other office.

"Small, but I'll take it. How much?"

Jodi was flabbergast at the man's arrogance. She hadn't offered to rent it yet. She started to protest, then realized her true economic plight. She was in no position to be picky. Besides, if he'd been sent by Senator Trudeau, he must be acceptable.

"You know Senator Trudeau?"

"Oh yes," said Tony, relaxing into wide grin. "She's our Senator. In Nevada we all love her. A gal with real class. She's a great friend of the boss."

"Real friendly, eh?"

"Well, I wouldn't say that," said Tony, scratching his head. "I've seen them standing toe-to-toe in some real brawls. But they respect each other. When it's all over they have a big laugh."

Jodi made a sudden decision.

"Okay. You can have the other room for half my rent. I'll have the filing cabinet moved out tomorrow."

"Don't fret. I'll move it myself. Where do you want it?"

"Stick it in a corner of the reception room."

After Tony Pellini left Jodi sat behind her desk and wondered at her sanity. She hadn't checked Tony's references or even called Senator Trudeau to see if he was telling the truth. And in the morning she would have a totally strange man in the office who had the manners of a wild boar. Financial desperation can create strange bedfellows. She decided to visit the Times newspaper morgue and learn more about Guy Ericson. If he really owned a professional football team and a movie studio, there were bound to be articles.

Also there might be some ancillary information about Tony Pellini.

There was.

From the news morgue Jodi learned Tony Pellini had been a middle-weight boxer who fought many times in the Palace Casino. Later he had dragged Guy Ericson from a flaming airplane which crash landed at the airport. Taking a job as bodyguard after he quit boxing, Pellini was indicted for killing a man with his fists. He won the case. The man had been a member of the Mafia trying to intimidate Guy Ericson. One reason the jury exonerated him was that Nevada politicians, with Guy's backing, came to Pellini's support. If the Mafia managed to crack into Casino gambling, the state's major source of income would be forever jeopardized.

Jodi learned Guy Ericson was a wheel in Nevada. Tony had been speaking the truth. She was also grateful Tony had insisted on paying three months' rent in advance.

"Don't want to be bothered with damn paperwork every time I take a deep breath," Tony said.

Jodi had trouble fathoming Tony Pellini. He was an odd bloke. Quiet and played his cards close to the vest. Wringing information out of Pellini was like wringing a dry mop. When he did say something, it was often brutally blunt. He seemed insensitive to the feelings of others. Yet Jodi was to learn it was a protective shell developed in the fight game.

After several weeks of mysterious comings and goings in the office, Tony plopped himself in the single chair opposite Jodi's desk.

"Got any clients yet?"

Jodi's face flushed with embarrassment. She thought the question impertinent until she noted the concerned look on Tony's face.

"Well no, not exactly," she stammered, brushing away a stray lock of unruly hair.

"Not exactly," Tony retorted. "What the hell's that mean? Yes or no?"

Jodi sat up in her chair and looked Pellini straight in the eyes. She refused to be intimidated.

"It means no. Clients want ex-congressmen or staff chairmen. It's not because I haven't tried."

Jodi thought of the months she had pounded the halls of congress and industry. She even had taken to making the rounds in tennis shoes to ease her aching feet. Then, just before an appointment, she would dart into a ladies' room and take out a pair of high heels from her oversized handbag and wear them into the interview.

"How about running some errands for me? Boss wants me to hire a runner." Pellini grinned. "Put those good looking legs to work. Looks like you could use the money while you're waiting to snare that big client. Pays $150 per day."

"I don't know," Jodi hesitated. His offer sounded a bit like charity.

"Doesn't equal the $500 per hour lawyers in this town charge. But boss said I could probably use a young man or grab a page from the hill." Pellini grinned, scouring Jodi's figure boldly with his eyes. "With the right clothes you could pass as a small boy."

Jodi didn't know whether to be angry or thankful until she saw the rare twinkle in Tony's dark eyes.

"I could use the money," she said, throwing pride to the winds. "What would I have to do?"

"The boss," Tony rarely referred to Guy Ericson by name, "has publicity people turning out reams fighting the new budget bill. It's got a lousy clause penalizing gambling and sports. He wants releases delivered in person to all members of congress and their staffs."

"Why not send them by mail?"

"Don't be stupid. They'd only end in waste baskets."

Jodi ignored Tony's abrasive remark.

"I'll give it a try. At least until I find a client. No strings attached?"

"No strings attached," said Tony. A flicker of a smile danced at the corners of his mouth. "You're a suspicious tyke. They wean you on espionage in Nebraska?"

Jodi laughed heartily at Tony's sour expression.

"No. My father and grandfather were horse traders. They taught me not to get suckered by another Abraham Lincoln?"

"Abraham Lincoln? Old honest Abe? I don't get it?"

"Abraham Lincoln agreed to trade horses sight unseen with a known con artist."

"So?"

"Everyone thought him a sucker and advised against it."

"But he went ahead anyway?"

"Yep. Abe traded the rascal a wooden saw horse."

Tony's mouth spread into a wide grin. He chuckled through his nose, pulling out a handkerchief.

"The tricky bastard," he nodded approvingly. "We could sure use Lincoln in the White House today."

"When do I start?"

"Tomorrow. Work only until one o'clock. That should leave time for you to pursue your own interests. A deal?"

"A deal...and thanks. Now I'll be able to eat better."

"One more thing."

"Yes?"

"How about dinner this evening? You don't look like you've had a square meal in a month."

Jodi stiffened. "I've some things..."

Tony interrupted.

"There goes devious Abe again. You've got no plans. Anyhow, you're not my type. I like grateful older rich women. You strike out on all counts. I spent enough years in training camps. I don't relish starting over with a beautiful but broke and useless young filly. Since you've not lined up any boy friends willing to spring, a good dinner might lower your atmospheric pressure."

"I'd be delighted," Jodi said, ashamed of her reluctance to trust anyone. Her affair in the Speaker's office had made her paranoid.

From that moment on, Tony Pellini practically moved in as Jodi's manager and protector. She advanced from errand boy to handling legal briefs for Guy Ericson, although she had yet to meet him.

"Soon as you're ready, I'll take you to Vegas," promised Tony. "Boss a great one but a finicky hombre. Stickler for details. Got a stable of lawyers but gets rid of them like quarterbacks or horses when they don't perform."

Jodi learned Guy Ericson used Tony like the Godfather used the executioner. He was right arm to the boss. His assignment in Washington was only until the budget war was over, which should end in another two or three weeks. Meanwhile Tony flew out to Vegas every other week-end, arriving back on the overnight Red Eye early Tuesday morning.

She knew Tony slipped untold amounts of money to the proper sources but she never mentioned the matter. One morning she found a partially opened envelope with $10,000 on his desk as she scurried to answer his telephone which he had forgotten to place on recording. Also, next time she tried to pay her rent she found it had been paid for one full year. Arguing with Tony was fruitless. Occasionally he could resort to pure tyranny.

One Tuesday noon Tony stormed into the office, eyes bloodshot from the overnight trip, and threw his briefcase against the wall.

"Goddamned morons never seem to learn," he yelled.

Jodi rushed in to fathom the touble. She had never seen Tony lose his temper.

"A problem?" she ventured.

Tony threw himself into a chair, catching himself to prevent an upset.

"Boss gave me hell. As if I could control the flakes."

Jodi stood silent, waiting.

"Boss owns thousands of acres. EPA's been yelling 'fire' again. Like Justice Holmes said, yelling fire in a crowded theatre is not protected by the First Amendment but the nuts think otherwise."

"What nuts?"

"Media. Congress. Environmental kooks." Tony hit the desk with his fist. "Nuts who make a habit of yelling fire."

Jodi surmised Tony's explosion had something to do with Guy Ericson's land. His voice rose higher and higher.

"They yelled fire and got Hollywood involved on the Apple Adar fiasco.

"They yelled fire on the ozone layer...

"They yelled fire on earth warming...

"They yelled fire on the snail darter, now ain't that a dilly?

"They yelled fire on energy under President Carter, creating panic and gas lines when the world was awash in oil...

"They yelled fire on atomic waste to create a superfund of billions for nuclear waste extracted from the gullible taxpayer. Ninety percent of which went to lawyers.

"Labor department egg heads yelled fire in order to retrain military unemployed so they can become industry unemployed...

"Now the same mental midgets are yelling fire on wetlands, trying to force conversion of good land into wetlands."

"Anything I can do?" Jodi offered tentatively, hoping to calm Tony.

He sat silently, meditatively, then looked up with a wry smile.

"Yeah. Get busy on some legal briefs against EPA. Time we gave them as much trouble as they're giving us. Shouldn't be too difficult. They're hiring ooze from the bottom of law school barrels. Zealots who don't possess one grain of common sense."

5. THE PUB RESTAURANT

"Ever eaten here before?"

Tony Pellini's tone was amused as the waiter fawned over the table. The two were having lunch at the PUB, a restaurant located within two blocks of the U.S. Capitol.

"Once. With staff, when I worked for the Speaker. This is a favorite hangout of Speaker Metzer and his buddies."

Jodi blushed with embarrassment as she noted the attention bestowed upon Tony. While she'd never heard of Guy Ericson or Tony Pellini before he called, it seems the world knew who they were. Small wonder Tony wanted to know if the only crop in Nebraska was weeds.

"I've never asked you about your stint with the Speaker, since you seemed reluctant to discuss it," said Tony, watching Jodi over the top of his drink. "Feel up to it now?"

"Not one of my life's favorite episodes," said Jodi.

She wondered? Why did people never tire hearing about the time you got thrown from a horse and broke your arm. But the fact she won highly competitive horse races and a tough calf-roping contest was a bore.

She had been voted Beauty Queen in University and knocked her brains out to rank 15th in a law class of 387. Nothing could be of less interest. Now Tony wanted to know about the most devastating experience in her life--attempted rape and firing from the best job she ever had. Though she had come to respect him like a brother she felt like rapping him on the head with the wine bottle.

Tony was quick to notice Jodi's reluctance.

"Forget I asked. A stupid question. Didn't mean to pry. Concerned your isolation's not healthy."

Jodi frowned. "I'll clue you in on the whole story one of these days. Right now I'm still so enraged with the Speaker I'm practically incoherent when his name comes up."

"Skip it." Tony glanced down at the menu. "The crab cakes are good here."

At the Maitre d's stand there was a slight commotion.

"Oh, oh. Speak of the devil," said Tony nodding towards the entrance."

Speaker "Fat Albert" Metzer stood impatiently looking over the room with his party of five. Following closely was "Jam" Johnson, TV talking head on Sunday's full hour news program; Ali "Knuckles" Rahman, the Speaker's Administrative Assistant; the Reverend Jesse Rowan, head of the black caucus and Chairman of the House Judiciary Committee; and Senate Majority Leader Mordecai Meyer, inseparable buddy of the Speaker.

The hostess led the group directly by Tony Pellini's table towards one in the far corner of the room. In passing, the Speaker stopped.

"Glad to see you visiting Washington, Tony. Give my best to your boss. I plan to visit his track during the holidays." The Speaker was an avid horse player seen often at the major tracks.

Without rising, Tony acknowledged the Speaker's greeting.

"Perhaps Miss Zukov and I will see you there?" he said as the Speaker moved on.

Jodi flashed an appreciative grin when the party was out of earshot.

"Thanks for the plug. I realize you wanted to reprimand the Speaker for ignoring me. He acted like this was an empty chair. I'm more likely to visit Rome than any track in California."

"Clever rascal, aren't you?" said Tony. "That's exactly what I wanted to do."

"He hates me."

"So it won't hurt to mix a little fear into that hate. My boss scares him to death."

"Why?"

"Boss owns control of Material Resources International, Inc., a conglomerate in aircraft, munitions, mineral holdings, communications and energy. The company has a big U.S. contract employing Russians to defuse atomic weapons in former Soviet Union republics."

"How in the world does that affect the Speaker?"

"Good old 'Fat Albert' has a pot full of Iranians in his district. Mosques literally dot the landscape. At one time they showered him with money. President Roland changed all that when he tightened the screws through MRI on Iranian weapon buys from old-line communists. Shiite Muslims want the Speaker to do something about it."

"Getting the President to change his policy against Iran? About as likely as getting my old job back."

Tony smiled indulgently.

"Metzer wants MRI to allow the sale of old Soviet weapons to Iran before they're completely dismantled. Also wants the company to release former top Soviet scientists so they can work in Iran."

"Where they'll be paid by Iranian oil money?"

"Yeah. But boss shares the President's dislike of Iranian fanatics. Barely escaped with his life across the northern border when Khomeini grabbed power. Trudged for miles with bleeding feet over the mountains until rescued by Turks."

"My ancestors came from Russia. They always considered Persians treacherous, way back to the days of Genghis Khan."

"Notice you carry the instincts of a Genghis Khan," Tony said slyly.

"My dad taught me there was a lot more to Genghis than the accepted folklore of his savagery."

"For instance?"

"Aside from the fact he had no use for his enemies, Genghis was a surprisingly enlightened and liberal ruler."

"Hard to believe?"

"Torture was rare and forbidden--far less cruel than the laws of Islam. There was no discrimination between subject peoples. Even conquered women were emancipated and made equal with Mongol women.. His was the first empire to insist on religious freedom--allowing churches, mosques and temples to stand side by side. Among the few crimes punishable by death were adultery and cattle theft.

"The Khan was a shrewd judge of men," Jodi continued. "Men elected their own officers in the lower ranks and were promoted for ability and achievement alone. Discipline was incredible but not harsh. Always outnumbered, Genghis employed a highly sophisticated intelligence network, advancing armies on a wide front with highly developed systems of communication."

Tony smiled. "Sounds like a brain wash from your dad?"

Jodi shook her head.

"Nope. Facts are facts. Without question his army was the best army in the world. He permitted no infighting. Communities, once rebuilt, thrived under secure and protected tariffs, trade routes, farms and markets. He suppressed customs which preserved poverty and ethnic discord. He forbade any man to own a slave, spread groups among different units of the army to inspire cohesion and disbanded feudal systems under which the right to command depended upon the extent of a man's estates.

"But basic to everything," said Jodi, recalling her father's stories, "was that all men in the Mongol Empire over the age of twenty--with few exceptions--were liable for military service. He built unity which lasted hundreds of years after his death--longer than Napoleon, Alexander the Great or even America thus far. General Schwarzkopf's famous flanking movement in Iraq was old hat to Genghis and his top commander Subedei, who performed the maneuver repeatedly."

"Okay," laughed Tony. "I concede. Genghis was quite a guy."

Jodi shook her head. "I only wish some of our leaders could learn from history. We've learned little on how to solve our problems. Social programs and sticking tentative toes into the water instead of throwing the whole body into the pool. That's how dad taught me to swim."

"Like Genghis? Include everybody?"

"Roger. Half-ass measures never work."

"Fundamental Shiite Iranians, your former Persians, again are becoming dangerous," said Tony. "They're trying to build a war machine to control the Middle East. Found it's a hell of a lot easier to simply buy weapons and scientists than start from scratch."

Tony was a great deal more knowledgeable than an ex-boxer.

"Where'd you learn all this?" Jodi questioned slyly.

"Graduated from University too. You never bothered to ask. Got started as boxing champion at University of Nevada in Las Vegas. Surprised?"

A chagrined Jodi lowered her eyelashes in embarrassment.

"I always took you for..." she stammered.

"A hoodlum?" Tony laughed. "Well, I practically became one until rescued by Ericson. Boxing leads one down that trail."

"Until you joined Guy Ericson?"

"Yeah. He hired me after I rescued him from an airplane crash and pushed me into his business ventures."

"Why would the U.S. give your boss a big contract to hire Russians when so many Americans need work?"

"Top scientists inside Russia," corrected Tony. "Not ordinary Russians."

Various people stopped by the table to pay their respects to Pellini, always with the admonition to give their best to Guy Ericson. Representatives, senators, business executives and sports enthusiasts. Jodi finally was beginning to understand how much influence Guy Ericson wielded.

"What's the matter with sending our own scientists? It's our money?"

"We do. A few to monitor and control operations. But if former Soviet atomic scientists can't earn a good living, they're ready prey for Iran, North Korea, China or any other renegade country that wants to hire their knowledge of sophisticated weapons.."

"Makes sense," said Jodi. "Preventive medicine. Devil of a sight cheaper than another cold war. Russia's a huge country and hasn't settled down yet."

"Job will take years. Meanwhile our dismantling supervision provides better infiltration than the CIA."

Jodi stared at Tony in disbelief.

"Your boss also in the spy business?"

Tony chuckled. "He practically operates a separate spy network. Knows where and how the most dangerous weapons are maintained. Material Resources has caught and blocked the secret sale of millions in arms to Iran."

"Does President Roland know what's going on?"

"He's behind the whole thing. Gave my boss the contract. Speaker fought the operation tooth and nail."

"Spying never seems to end. Remember the double agent and his wife who worked in the CIA for nine years?"

"Aldrich Ames' monkey wrench nearly wrecked Clinton and Yeltsin's pact of brotherly love," laughed Tony.

"Speaker Metzer's Administrative Assistant is a half-ass spy in my opinion," said Jodi. "His father's an Ayatollah with direct connections to Iran."

"Got a nest of them in 'Fat Albert's' district."

"Guess I'm in pretty good company when the Speaker hates both the President and myself?"

"Exactly what I've been trying to tell you," laughed Pellini. "Stay away from Iranians. They take hatred of the infidel to fanatic ends."

Jodi had the unforgiving, depthless quality of the single minded Mongol warrior of her ancestry. Mongols had a deep-seated hatred of Iranian treachery. A one-time traitor was never forgiven. A broken word never forgotten. Objectives were plain and simple--win. Genghis Khan had been her grandfather and father's idol. They filled her younger years with tales of his victories and how he obtained them against incredible numerical odds. A foe who jumped sides to surrender was immediately executed.

Jodi looked up as a bottle of champagne arrived at the table, compliments of the management.

"To keep his seat the Speaker plays a dangerous game with voters," said Jodi. "He's striding two galloping horses. I've seen it done at rodeos."

"No problem. Long as they're harnessed together and going the same way. But his Muslims and the U.S. are headed for a wide fork in the road."

Jodi gave Tony a slow satisfying smile, thinking of the possibilities.

"I'd like to be there when the Speaker tries to do the splits."

A disturbed frown creased Tony's forehead.

"Trouble is, most Americans are prone to pay no attention until a Pearl Harbor hits. Yet Shiite fundamentalists take secret oaths from their ayatollahs to kill infidels who thwart their actions."

Jodi ran her tongue nervously over her lower lip.

"Doesn't your boss and the President know this?"

"You bet. They've learned to be careful and live with it."

Jodi and Tony had finished their lunch when Tony jumped ecstatically to his feet. With her back to the Maitre 'd, Jodi had no idea why. Tony normally wouldn't jump to attention for a queen. She turned her head in time to catch a wave from Senator May Trudeau. She was accompanied by General Pete Hardin, in civilian clothes. General Hardin rarely wore his uniform outside the Pentagon, despite being Air Force Chief of Staff.

Tony pulled back a chair for Senator Trudeau.

"Please join us. Jodi and I are having a three-martini lunch without the martinis."

Senator Trudeau glanced at the champagne bucket.

"Must be some kind of celebration?"

"Compliments of the house," said Tony. He raised his hand and the Captain was at his side immediately. "Another bottle please." He turned to the others. "Can't have guests treated unequally, can we General?"

From the way Tony addressed the group, Jodi realized they were old friends.

Senator Trudeau was the femme fatale of the U.S. Senate. Her dark hair, loosely secured in back with a bow of two white carnations, dropped to her shoulders. A bright yellow blouse and dark green skirt were secured at the waist by a wide leather belt.

The media seemed to need a Senator Trudeau in the U.S. Capitol each generation. If not the President's wife, such as a Dolly Madison, Jackie Kennedy or Hillary Clinton, then someone from the House or Senate.

Senator Trudeau quipped to Jodi.

"Tony behaving himself? If not, let me know. I've got an open line to his boss."

"With Tony around, it's like I'm back in Sunday school," said Jodi. "He's got more eyes around this town than the FBI."

May glanced about the room as she waited for the Captain to pour from the new bottle of champagne.

"I see we've got prestigious company. Members from both houses plus the Speaker's favorite leak to the media."

"Must be planning to slaughter some poor victim," said General Hardin. He'd had a bitter grueling in his appointment to Air Force Chief.

"With 'Jam' Johnson poised for another scoop," added May.

Jodi took a sip of champagne, glancing carefully at General Hardin. Her heart skipped a beat. He was frightfully good looking with an easy grin, deep dimples and a lean tanned face. She had seen him from a distance but up close his magnetism tugged at her from across the table. She knew Pete was May's special capture and tolerated no intrusions. Both were good friends to have. General Hardin had built quite a reputation as a young colonel in Desert Storm.

"They'd love to torpedo President Roland but he's a tough hombre from two wars--Viet Nam and Desert Storm," said General Hardin.

"The Speaker's careful to feed his own ducks," said Senator Trudeau, staring disdainfully at the corner table. "Controls the black caucus through the Reverend Jesse Rowan."

"First time I've seen Jesse Rowan without a megaphone in his hand," laughed General Hardin.

Tony turned to Jodi with a wide smile.

"See. You're not the only one who thinks the Speaker's a buffoon."

A few minutes later the Speaker's crew got up to leave. As they filed past the table making pleasant perfunctory remarks, Ali 'Knuckles' Rahman lingered behind. He stopped opposite Tony, staring down from his huge bulk. To Jodi he looked even more menacing than when he had attacked her.

"You surprise me, Tony," said a grinning Rahman.

"How's that, Ali?" Tony said in cordial tones.

"Bringing a tramp and whore to be seen with you in a public place. Has she paid her tax? Whores have to pay a license tax in the District."

Jodi gasped. She took the champagne glass and flung its contents full in Rahman's face.

He smiled as he wiped the stinging champagne out of his eyes.

"A tramp and her manners never part company," he said as he scurried to the exit.

Jodi furiously rose from her seat before Tony laid a restraining hand on her arm, his eyes glowering with suppressed rage.

"Never get mad, get even. A day will come."

The group fell into an embarrassed silence as tears filled Jodi's eyes while she fought to gain control. Tony finally ventured.

"I'm sure you'd feel better if you'd tell us what that was all about?"

Jodi nodded agreement. "I hadn't wanted to, but Ali's animal behavior makes an explanation imperative."

Jodi poured out the whole story of the attempted rape, how she had sent Rahman to the hospital and her firing from the Speaker's office. Feeling she was among friends she left out none of the lurid details.

"You should bring charges," said Senator Trudeau indignantly. "That animal needs to be caged."

"No," said Jodi. "I've thought a great deal about that. The victim is forever tainted. Lawyers and the media would twist such a high profile event into a circus. Part of the public would always wonder...is she really a whore?"

An angry Senator Trudeau said testily.

"Rahman shouldn't get off Scott free to humiliate you again."

"Don't worry," said Jodi, biting her lower lip until she nearly drew blood. "My time will come. You can depend on it."

Tony glanced at the resolute beauty sitting beside him and had no doubts. He could have stood up and decked the big oaf but years in the ring had taught him control was a more powerful weapon than rage. Also he'd had one killing with his fists in his background.. For an ex-professional fighter to hit an amateur, no matter how big, was a felony. He resolved Ali was living on borrowed time.

6. ALI 'KNUCKLES' RAHMAN

Jodi Zukov knew a great deal about Ali Rahman's habits. He wasn't secretive about them, announcing loud and clear the mistakes of Redskins' football or the Washington Bullets' basketball teams. He religiously attended the games.

Ali lived in an apartment complex near the stadium. Not particularly a savory neighborhood but away from the social elite his father despised. He got twice the apartment size and underground parking for half the price of one near the capitol. There was little to fear for a 300 pound ex-professional football player with his reputation. Neighborhood hoodlums left him alone.

Jodi visited her office rarely in the week following her disastrous encounter with Ali Rahman. Tony was concerned but left to spend time in California. He felt badly that he couldn't offer her a shoulder to cry on.

Jodi had her own agenda. She resurrected old paraphernalia from a trunk buried in her closet. She nostalgically dragged out rodeo and ranch items, sitting on the floor and examining them one by one. She had almost decided to call it quits and return home to work the ranch. Trouble was, following her father's sudden death from stress, it barely eked out a living for the tenants after taxes and legal bills were paid. Government regulatory action had stripped the cupboards bare as well as cause the untimely death of her father.

Jodi sat toying with an old block and tackle used for lifting bales of hay into the barn. It also was used to raise horses to their feet to prevent lung collapse when sick. With block and tackle it was possible to lift heavy loads by the application of only a relatively small force. She'd had new ropes put on the pulleys in Cheyenne and the repair shop had sent them to her in Washington instead of the ranch. As she idly tested the blocks, she suddenly had an inspiration.

* * *

Ali Rahman drove into the dimly lighted underground garage just before midnight. He whistled merrily. The Bullets had defeated the Celtics in a last-minute surge. A basketball rarity. He closed the car door and strode with a jaunty step towards the building's rickety elevator.

Without warning a rope dropped over his shoulders. He'd been lassoed and his hands pinned to his sides.

"All right, who's the jokester," he said irritably, attempting to slip out from under the rope.

Abruptly he was jerked off his feet with his body swinging and twisting in the air. Before he could make out his assailant he felt his feet being lashed together. An expert at calf roping finished the job in less than ten seconds. Next his upper body was secured by a third rope.

Jodi stepped back to observe her handiwork.

"Wriggle out of that shoulder rope and you'll catapult head first into the concrete and smash that delicate brain of yours into a bloody pulp," she said as she anchored the swinging body to a nearby concrete pillar.

"Oh, it's you," snarled Ali Rahman. "You'll live to regret this, you stupid bitch."

Jodi calmly walked around the helpless Rahman, testing the ropes as she did at all rodeos.

"Now what kind of tramp am I?"

"A bitch and a whore," cried an enraged Rahman, twisting and turning with his feet off the floor. His face turned purple as he struggled to burst out of his rope prison. The rope to the pillar finally stopped his wild swinging.

Jodi smiled as she swung around to the front. Easier than roping and securing a wild 300 pound maverick in a rodeo.

"Now I'm going to ask you one more time. What kind of a lady am I?"

"A bitch and a whore," screamed Rahman, beginning to feel the pain from the rope holding him aloft. But he'd endured worse pain in professional football. The little bitch wasn't going to get the best of him. His struggles only made the rope bite deeper.

"My, what a nasty man you are," said Jodi, looking up into his eyes.

"You'll live to regret this, you little fool." Rahman hunched his huge shoulders in one last effort to spring free.

"May I ask why?" Jodi spoke with exaggerated courtesy.

"I'll see that you spend the rest of your life in prison."

"Before I go," said a suddenly grim faced Jodi, "I'm going to teach a fat pig some manners."

She unhurriedly walked to a nearby car and took a tire iron from an opened trunk. Without another word she returned to face the dangling Rahman. For the first time the reality of his predicament dawned on Ali. This was no prank.

"Good god," he cried, his face ashen.

Jodi swung the iron hard against Rahman's right shin. The crack of steel against bone echoed around the garage.

He screamed from the pain. "For chrissake you've smashed my leg."

Jodi stared at the dangling giant.

"I've heard no apology. For such a smart ass you're a slow learner."

She brought the tire iron back around her waist and swung down hard on his other leg.

Ali again screamed in tortured agony. He started to whimper.

"Oh my God, you've broken both legs."

"Next it's your mouth," said Jodi, swinging the iron inches across the front of his face. "That's where most of the trouble starts. Perhaps if I crushed your teeth back into your throat in bits and pieces I could shut off the source of your foul remarks." She made a move as if to execute the threat.

"Please...please..." Tears started to roll down Rahman's cheeks.

"I'm listening," was all a grim-faced Jodi had to say.

She smashed the tool against Ali's collar bone, snapping it with an explosive pop.

"I can't hear you," she said.

"You're a lady," Rahman screamed, his voice cracking into hoarse sobs. "Please to God. You're a divine lady."

"To you I'll always be a lady...right?"

"Always," Rahman moaned in unspeakable agony. "Always...." His voice trailed off.

Suddenly he went limp, passing out from the pain.

Jodi stood watching the limp figure for a full minute. She returned the tire iron to the trunk of the car, then carefully lowered Rahman's body to the floor. After unwrapping the ropes and retrieving the pulley and blocks she covered Ali Rahman with a blanket to prevent shock. Next, she called an ambulance on 911 from her car phone, slowly driving out of the deserted garage.

Not until she watched the ambulance enter the garage did she leave her parking place across the street and drive home.

An afternoon edition the next day carried a front-page story about four thugs jumping the Speaker's aide and beating him with a heavy weapon. The hospital said he would have a full recovery but it would take time.

The police testified that Ali Rahman's description of the men, who wore masks, was so vague they had little to go on.

A fearless professional football player with a national image could admit only to being over-powered by no less than four men, never by a snip of a woman.

7. GUY ERICSON

"The last thing I need around here is a female lawyer. They're big trouble."

Guy Ericson put his feet on the desk and glared at Tony Pellini.

"You ought to know better."

"She's done the company a lot of favors. Her damn good looks and plain guts got her past doors I couldn't crack."

Guy Ericson lips curled in derision.

"She got paid, didn't she? Look, we've got more fancy looking broads per square foot in Vegas than they've got in the entire District of Columbia."

"I'm not asking you to hire Jodi Zukov," said Tony stubbornly. "Only that you reward her work with a trip out here." He added sheepishly, "I half-way promised her."

A wide grin crossed Guy Ericson's handsome features.

"Personal eh? Why the hell didn't you say so in the first place? Bring her out. Set her up in a company suite. But she's your baby to handle. I've got enough broads tripping over my feet to start a Turkish harem."

Tony decided there was no use trying to explain to his over-sexed boss there was absolutely nothing more between Jodi and him than between an older brother and sister.

"Kinda gone berserk since the wife died?" Tony needled.

Guy and Tony were often brutally frank with each other. Guy Ericson normally liked the infighting with Pellini. He was the only one of his employees who would stand up and counter-punch blow-for-blow. Most of the others, fearful of Ericson's volatile temper, found it wiser to brown-nose the boss.

"Really none of your damn business," retorted Ericson in an unpleasant tone.

"Just worried about your staying power, boss," said an unperturbed Tony. "I hate looking for work."

Tony admired and respected his boss. But Guy Ericson in Vegas, after the death of his wife two years ago, was like dropping a cache of honey into a field of flies. Rich. Single. Tall. Blonde with a ready smile, he was a *catch.* A dedicated athlete who carried a two handicap in golf, Guy was offered a tryout with a pro basketball team when he graduated from University. He never made it. Six foot three is a nice civilian size but a dwarf on the professional circuit.

Guy was Norwegian. His grandfather migrated from South Dakota in the early days when Las Vegas first started to become a boom town. His grandfather bought cheap land and pulled together a few local companies but his father went hog wild, using the junk bond tactics of Michael Milliken to build a conglomerate in Nevada and Los Angeles. When his father died from a heart attack, brought on by sleepless nights and an unremitting squander of priceless energy, Guy rejected Wall Street and went after the fast-moving exciting market of sports, gambling, airplanes, and international risks. The latest of these was Material Resources International, Inc.

Guy wore a tailored blue jacket, cream colored trousers and shirt with an open collar. He considered the tie a Machiavellian harness designed by women to control men. Something like a handy leash.

"Your gal may be everything you say, Tony," said Guy, relenting his harsh demeanor. He got up and walked over to put a hand on Tony's shoulder. "But dump a woman lawyer into my nest of sharks? Never. Like Ahab, I like to harpoon them myself. Women just aren't aggressive enough."

"Boss, I'm not believing my ears. Surrounded by women you say something so silly. You ought to visit Washington."

"Never again, if I live to be a hundred."

"You might learn about aggressive women. In congress and in the feminine movement are women who would leave Jesse Owens on the starting blocks."

"You say Jodi Zukov helped the company," said Ericson with an annoyed frown. "Then why the hell we getting all this flack about MRI's contract to defuse the atomic warheads in Ukraine, Kazakhstan, Belarus and Russia?"

"We delivered all your releases...talked to the principals. But the Speaker controls congress and he's at loggerheads with the President over Iran. Remember, it's the President, not the Speaker, who gave MRI the contract in the first place."

"Yeah, I know." Guy's voice was quiet and bitter. "A lion and a hyena. The Speaker wanted to be on the ticket for Vice President--lord knows why. President Roland kept him off in favor of Whitaker, a war veteran."

"After Clinton and his leper appointments, the public had enough freaks," said Tony. "Then followed the Republican congress. Now an old warhorse has captured the Speaker's chair and his underlying real beef is the way the White House and you lowered the boom on Iran."

Ericson got up and paced nervously behind his chair.

"No question. Through MRI the President's come down hard on Iran. And the Speaker's fought him every step of the way."

"This Congress wants Iran's oil money to recapture obsolete social programs and they don't give a damn if they arm the devil to get it. On top of that, planking out U.S. money to defuse Russian warheads is like prodding a wounded wild boar with a sharp stick," laughed Tony. "Each time the subject came up when I was back in Washington, the Speaker went berserk with rage."

Guy Ericson's genial humor was gone. He thought of the many headaches congressmen were creating for his working scientists overseas. Their constant trips, invasions into delicate operations, second-guessing opinions on sophisticated technically demanding tasks, all added to delay and increased costs--to which the same congressmen then objected. Dismantling atomic weapons was not a job to be supervised by a debating team.

Congress by pouring billions into the Department of Energy, ninety percent of which had gone to lawyers, had micromanaged its own atomic wastes into the biggest financial boondoggle in U.S. history. Now it wanted to duplicate its failure overseas.

*Modern strategic warheads consist of a primary fission explosive and a thermonuclear (*fusion*) gut that is ignited by the explosive. The hollow, spherical gut of the primary holds the warhead's plutonium, generally three to four kilograms. Surplus Soviet warheads, including over 10,000 already dismantled, contain about 100 tons of plutonium and more than 500 tons of HEU--highly enriched uranium.*

Considering that three pounds of plutonium can destroy a city, Guy Ericson felt his scientists didn't need oversight by politicians with megaphones.

"I doubt half congress got more than high school math or physics," muttered Ericson. "They took political science, economics, business administration or law. They know as much about a nuclear warhead as an aborigine."

Tony sat silent. He didn't know any more about nuclear weaponry than congress.

Guy Ericson shook his head angrily.

"The Speaker wants MRI to allow the sale of recycled plutonium. Such stuff readily can be turned back into deadly smaller weapons by a renegade country like Iran."

"We really need to defuse congress," said Tony.

"Idiots don't seem to comprehend the old Soviet Union had a tough system of border controls. Now 15 independent states and 5,500 miles of new frontiers beckon with open arms to smugglers."

"To say nothing of the rampant corruption," said Tony, repeating what he had learned from talks with friends in the CIA. "Old Soviet party bosses still control the vast majority of state resources and assets. Smuggled money keeps the old guard active."

"Corruption is so bad the FBI had to establish agents in Russia," said Ericson, rolling back his chair and springing to his feet.

Tony watched his boss pace the room in silence for a full minute before speaking.

"The FBI worries a new Mafia has developed in the chaotic republics of the old Soviet empire and will soon move into the U.S. Just as the Sicilians did after Mussolini left the nation destitute."

"This Mafia would be far more dangerous."

"How?"

"By peddling weapons of annihilation instead of drugs. The old empire is loaded with dangerous arms ripe for plunder."

Tony nodded agreement.

"Iran's mullahs blanket the four atomic republics buying everything they can bribe or steal."

"The Speaker thinks he can conquer the ayatollahs with an olive branch," said Ericson. "The President knows better."

"Appeasing Iran is like appeasing any dictator. Whets their appetite and assures them they're dealing with wimps they've got buffaloed."

"Sadly Iran's right about United Nation wimps. They had only to look across nearby borders to the fate of Muslims at the hands of the Serbs."

"Yeah. The western world, including Europe and the United Nations, faced the Serbs less courageously than Chamberlain faced Hitler."

"The Bosnia war convinced Iran to open the throttle full speed in obtaining weapons they wanted. The Serbs proved the United Nations a body of chattering magpies. They got what they wanted long before the UN belatedly got its act together. The UN contributed to thousands getting brutally murdered."

Tony expressed his frustration at the way the world resists common-sense solutions.

"We're letting an evil state build a powerful military under our nose...with our full knowledge."

"The danger is down the line. They might produce a powerful leader like *Sulieman the Magnificent.*"

Guy Ericson returned to his desk. Politics annoyed him. He changed the subject.

"Well, bring your hot new delight out here. Meanwhile I've got these cables to answer." He looked up with a smile. "We'll have dinner together. I haven't seen you shill for a broad since that blonde ran off to Europe with a Frenchman."

He whistled softly. "And a lawyer no less. In the sack she can read you legal briefs."

Guy Ericson guffawed loudly as Tony spun on his heel and left the room.

8. LAS VEGAS

Jodi's tear-filled eyes glanced around the empty office. Soon the movers would come back upstairs for her empty desk and the chair she was sitting in. Her ten-month effort to break into the Washington scene had driven her on several occasions to contemplate suicide. It was a tough town. The only people who counted were on the ladder above you. Lawyers in Washington were like grains of sand on a rocky beach. The big financially successful firms unconscionably charged way too much while the failures percolated down into the mud and survived on scraps.

She had just made peace with the landlord and handed him the keys. Although the lease had a good six weeks to go Jodi wanted to get the ordeal behind her. She snapped up the first offer for the office furniture. The apartment was a different matter. It had been furnished, so there was no problem except with her clothes. Two large suitcases and a vanity case she kept with her, forwarding the rest to the ranch in Nebraska.

Jodi had no emotional qualms about leaving Washington. The place had been a complete disaster. Enemies seemed to abound everywhere. Aside from the few months of legal work provided by Tony Pellini and a few odd jobs tossed in her lap by Senator May Trudeau, her first year on her own had been a flop.

Industry wanted former congressmen or staffers. While nothing specific was ever mentioned--it never is in Washington--most of those in congress shunned her. The Speaker's office got out the word that Jodi was "flaky and big trouble." Leaks were more solid than facts to the Washington media. Presidents and Cabinet officers learned that to their dismay. Special Prosecutor Walsh had wrung $200 million out of the taxpayer with an unending litany of unproved leaks. The office was just another rip-off of taxpayer money.

Tony Pellini had gone to the boss about Jodi's predicament. He thought the boss might use another lawyer since he'd just fired a stable of five. But Tony didn't want Guy Ericson to know the depth of Jodi's travail. Men like Ericson thought all unemployment was due to incompetence. When Tony's pitch to hire Jodi as a lawyer met an unequivocal NO, he suddenly had been forced to put Jodi's trip on a personal basis. This the boss understood. Guy's knowing grin irritated Tony, since he now had the unpleasant task of explaining the changed ground rules to Jodi.

Tony fretted. A couple of ingrates. Since neither his boss nor Jodi seemed eager for a meeting, perhaps he should junk the whole idea.

* * *

Jodi was all prepared to leave for Nebraska when she received Tony Pellini's phone call.

"The boss wants you out here tomorrow." A white lie he could live with until things got straightened out.

"I can't possibly..."

"You can if you get off your pretty ass right now," Tony said. "I've checked. A flight on Delta leaves from Dulles at 9:30 in the morning. You gain three hours on a five-hour flight. Puts you here in time for lunch."

"What if the flight's full?"

"I've made reservations. You're traveling first class. Pick up your ticket at the airport."

"I need time to pack," she said, staring at her fully packed suitcases and an empty apartment.

"Wear a short light skirt. This is Vegas. Show off those spectacular gams. Remember, this is a town where show girls are a dime a dozen. You'll have competition."

"I'm a lawyer, not a show girl."

"No kidding. You must've had me fooled when you were bringing in all those clients." Jodi heard him chuckle in his peculiar snorting manner. "Be here at high noon. I'll meet you at the airport."

"Will the boss be with you?"

"To pick up some barnacle from Washington? I should say not. But we're having dinner together. I'll clue you in at lunch so don't eat on the plane."

"I need to know..." Jodi said into a dead line. Tony Pellini had hung up.

Life guards rescue persons drowning in water. Tony Pellini was rescuing someone drowning in failure. Or so Jodi hoped as she took her seat next to a window up front in the Boeing 747. She planned to sleep most of the flight, restore nervous energy she'd drained away almost to the point of exhaustion.

Fortune again was not on her side.

Plopping down beside her a middle-aged man in wire-rimmed glasses pushed his brief case under the seat in front of him with his feet. He turned to Jodi with a frown.

"Might as well have pushed all that paper work and statistics out the window," he said with a feeble attempt at humor. "My congressman nodded like a mechanical doll then sent me on a wild goose chase to five difference agencies, none of which was interested in my problems."

"Often seems that way," said Jodi, trying to be pleasant as she closed her eyes in an attempt to stifle further conversation.

The disgruntled man failed to notice. He was brimming with animosity and still kicking the briefcase into place. "I'm a small business man," he said, as if speaking to himself while he loosened his tie. "Spent a lot of money coming here. Horses get less run-a-round at the Kentucky Derby than I got from the damned bureaucrats...Oh, I'm sorry. Slip of the tongue," he apologized, glancing at Jodi for the first time.

"You asleep?"

Jodi opened her eyes to the man peering into her face. He was a large man with piercing blue eyes and the reddish complexion of a man who had spent his life in the sun without benefit of sunscreen. A small band of white skin creased the top of his forehead.

"No. Just resting."

"You're not with the government?"

"Heavens no. I'm fleeing from it."

"Me too. May I never waste my time coming to this wretched place again. They care more for dead dinosaurs and rattlesnakes than they do for small business."

"Rattlesnakes?"

"Name's Myron Halsted," he said, sticking out his hand. "Yeah, rattlesnakes. Judge in Maryland sentenced a man to prison for killing wildlife..."

"Rattlesnakes?"

"Claimed a rattlesnake was wildlife. Big as a truck tire, too."

"Hard to believe."

"Not if you're raised in the Washington legal asylum. Inmates develop a different system of thinking."

"No question," Jodi straightened in her seat. Obviously she was not going to get any shut-eye.

"When I was a kid we went rattlesnake hunting in Wyoming," said Halsted. "Found thousands of them in caves crawling all over each other. Wonder how that judge would react to a truckload of wildlife dumped onto his front porch one morning?"

Jodi smiled at the thought. She wondered if Mr. Halsted might be persuaded to try the Speaker's office.

"Most judges spring from District Attorney offices. Few seldom see wildlife except on television," she said.

Myron Halsted took a second hard look at the person beside him. She was an uncommonly beautiful woman. *Blue-*

green eyes which contrasted sharply with her soft tanned skin and black hair. An unusually delightfully fragrance enveloped her. She was what he would have called sexy in his youth but there was a no-nonsense air about her which contradicted her looks. But it was her hands which attracted his attention. Strong and lean and supple.

Myron Halsted took a gamble.

"You wouldn't happen to have worked on a ranch?"

Jodi's shocked reaction upset him.

"Matter fact, I did," she said. "I didn't realize the hayseed stuck out."

"Oh, no," he said, chagrined and flustered. "I meant it as a compliment. You are a most beautiful woman, so I knew you couldn't be from Washington. Yet I couldn't help noticing your hands--strong like those of a sculptor or horsewoman or jockey. My daughter had hands like that before she was killed." He lapsed into silence.

"Sorry."

"She wanted to be a jockey. A horse fell on her."

Jodi watched Halsted struggle with his emotions for a few moments before breaking into his memories.

"I am a ranch and farm girl from Nebraska, three miles north of the Kansas border where Jesse James and other outlaws once roamed," Jodi said quietly. "Like your daughter, I loved horses. My father and grandfather focused their lives around horses and in the horse trade. I rode a horse to school across fields to save half-hour for chores."

Jodi hoped conversation about horses might bring the cheer back into Halsted's face.

"I was a farmer until government regulations drove me off," said Halsted. "From my meager remains I started a small lawn mower repair shop. Built it into a fair sized operation, too," he added proudly.

"That why you were in Washington? To see the Small Business people?"

"Naw. They're useless as tits on a boar to legitimate small businesses. They're into social engineering. Ethnic and inner city crap where they just throw money away."

"Then why the trip?"

"To get tax relief. I've been working my ass off and I want the government to let me keep some extra sweat."

Jodi was sympathetic. Halsted echoed the cry she heard from her own people.

Halsted turned to Jodi, surprised to hear a supportive voice.

"First I had nine employees," he said. "We repaired an average 18 machines per day. I needed 16 repairs to break even. The extra two machines were profit. Business was good so I decided to hire another five people. We had the business. Then along came the government. An affirmative action plan from the Equal Employment Opportunity Commission said I needed to hire two minorities. I agreed, only there weren't two minorities at my edge of town. Next I was told to put in environmental controls and meet new consumer product safety regulations. They forced me to hire a man who just sat around drinking coffee and supervising safety. Finally they sent in a woman to change my paper work and reporting functions. The Labor Department needed more reports, as did a half-dozen other agencies in government. Every month there was something new. Instead of extra employees making me money they were costing me money. I had to borrow from the bank and pay interest."

"Does make one feel helpless," said Jodi. "Get any promise of relief?"

Halsted laughed bitterly. "You kidding?"

"Who'd you see?" Jodi persisted.

"Saw the Secretary of Labor and never listened to such gobble-de-gook in my entire life."

"A Rhodes scholar from Oxford?"

"A lot of mouth. I know half-dozen Rhodes scholars. A debating society. A damn good one only they don't learn one iota of physics, math, chemistry or military art."

"A debating society?"

"That's it. Those who went there told me. They take different sides of a question and argue. See how adept they can become at talking on their feet."

"Sounds like unnecessary education for our politicians," smiled Jodi.

"When the Secretary of Labor proudly told me he was a Rhodes scholar, I knew I might as well leave. There was no hope," Halsted said, the anger rising in his voice.

"Weren't you prejudging?"

"Absolutely not. The guy looks at you yet doesn't hear a word you're saying."

"You're sure?"

"He could hardly wait to give me an academic lecture on the need for harmony between management and labor. Assured me training and education would solve my problems. Sounded like a canned speech taken from some video tape."

Jodi laughed softly. "So you met the federal ***bureaucracy***?"

"I told the idiot I ***was*** management. That my labor had all the education they needed to fix lawn mowers. I told him I could hire more people who didn't need additional education or training if government would just get off my back...and that included the Department of Labor which was loading me up with a hundred unnecessary tasks."

"What'd he say?"

"Nothing. Waited patiently, watching a bird build a nest outside his window. Then he launched ahead as if he hadn't heard a word I said. Repeated his diatribe about society needing more training and education."

Halsted took a deep breath, removing his wire-rimmed glasses and wiping away the moisture.

"The guy sounded like he was on drugs playing a broken record," said Halsted. "If he hadn't been a cabinet member I'd thought I was back in school listening to sophomore economics 101."

"Didn't you get help or learn anything from your trip?

"Yeah. I intend to take up my own crusade back home."

"Which is?"

"Get the people to send Washington a message which even an illiterate dog can't miss."

Jodi smiled. "A Rhodes scholar dog?"

Halsted grinned at his friendly passenger. She had a sense of humor.

"Yeah. Get off our back. Cut spending. Trim or abolish zealot agencies. Get out of Washington opium dens where working stiffs aren't admitted. Visit the home folks and listen. Either that or we fire you like any incompetent employee."

"The Republican congress under Newt Gingrich got off to a good start," said Jodi. "But like many before them, once in Washington they got swamped by the establishment. Like trying to hold your head above quicksand."

9. LUNCH WITH TONY

Lunch with Tony Pellini was at a small local restaurant one mile off the strip. Tony was nervous as a cat on a hot tin roof. He couldn't sit still. He bit a hang nail until it bled.

Jodi decided to find out what was wrong.

"All right. Let's have it. What's bugging you? Got no job? That it?"

Tony took Jodi's hand from the table and placed it between his two, looking at her with pain in his eyes. He blurted.

"Boss wants no female lawyers. No matter how hard I tried."

Jodi caught her breath. Now she really stood between a rock and a hard place. No money. The last time she looked in her purse, at the airplane counter, she had thirty-seven dollars and some odd change. A thousand miles from Nebraska. For a moment her temper flared.

"Why the hell didn't you play straight with me, Tony? I could have skipped the trip out here."

At once she saw the crestfallen hurt on Tony's face. No doubt he had gone to bat for her. She patted the back of his hand.

"Forgive me, Tony. We both made the mistake of thinking Guy Ericson would rescue me from my own follies."

"I truly tried. Told Guy how much you helped the company in Washington."

"When it's me you really helped." Jodi turned her face to hide the moisture in her eyes. "I only managed to postpone the inevitable."

"That's not all," Tony's large brown eyes betrayed his anxiety.

"Yes?" Jodi sensed little else could be more damaging.

"We're having dinner together tonight and the boss thinks you're my paramour."

Jodi sat upright in her chair, facing Tony with an accusing stare.

"You're treading on my reputation, Tony. An almost unforgivable action. You know better. I love you like a brother but even brothers have bounds."

"I was arguing your case to come to Vegas." Tony pleaded for understanding. "I thought surely the boss would hire you on my recommendation. His flat no was a complete surprise."

"And I was already on my way?"

"Right. With a sudden inspiration on how to pay for the trip I let him think you were my girl friend."

"That was okay to Guy?" Jodi winced.

"Sure." Tony smiled for the first time. "That's something he understands. Human foibles."

"A legitimate lawyer he doesn't want?" Jodi shook her head. "Yet paying for a floozy is more than welcome?"

"We can play along until we find a better solution." A relieved Tony noticed the reluctant acceptance. "I've got you set up in a small suite at the Palace." To allay the quizzical expression on Jodi's face he added.

"Don't worry. You'll not be bothered. I have my own apartment."

Jodi returned to eating her shrimp salad and toasted roll. She was famished. She had taken Tony's advice and not eaten on the plane.

Alone in the suite, Jodi fitfully unpacked. She wondered how she could survive dinner with a bastard *like Guy Ericson. In addition to running a hugely successful sports and business empire, he obviously ran a personal harem on the side.*

Undoubtedly, Mr. Guy Ericson was a Dr. Jekyll and Mr. Hyde.

From a notice pasted inside the door she determined her suite was $2200 per day. Done in soft pastel colors of blue, yellow and rose, the suite was quiet and relaxing. The place must have been prepared for a queen.

Jodi canvassed the rooms in slow easy steps, breathing in all it's rich features before an officious clerk found out her true penurious state and relegated her to a broom closet. She chuckled at the thought of the expression on the man's face if she offered to pay for the suite with the thirty-seven dollars in her purse. So this was how the wealthy slept while they gambled away fortunes?

Jodi glanced at her wrist watch. Four o'clock. Tony said he would drop by at 7:30 to pick her up for dinner with Dr. Jekyll at eight. Time for a hot shower and a quick nap. She was dead tired and needed to recuperate. The day promised an even worse finish.

At six o'clock the telephone woke Jodi from a sound sleep.

"Yes, Tony." She struggled into a sitting position, placing the phone into a more comfortable position.

"Tony, hell," said a laughing woman's voice. "This is Senator May Trudeau. Better watch how you answer calls or people might get wrong ideas."

"Tony was the only one who knew I was here," protested Jodi, flustered.

"I know. I just talked to him. Said you two were having dinner tonight with my buddy, Guy Ericson."

"Your buddy?" Jodi started to make a comment, thought better of it. How could a sharp beautiful senator like May Trudeau have anything to do with a Guy Ericson? Suddenly she realized Ericson must be a heavy contributor to Senator Trudeau's campaigns. That was the name of the game in Washington, heavy hitters could do no wrong.

Senator Trudeau sensed Jodi's strained response.

"I take it you've not met Guy yet?"

"No."

"You know the old cliché. Don't judge a book by its cover. Guy's incorrigible. We've had our brawls. He's a straight arrow with a mind like a steel trap. An intolerant Norwegian who thinks women ought to skin and cook the reindeer he drags in the front door. Stand up to him and he's a softy."

Jodi noticed Senator Trudeau didn't say anything about his morals. Perhaps, like Tony, May was blinded to his harem tendencies.

"I'll do my best to follow your advice," said Jodi, taking May's opinion with a grain of salt.

"Anyhow, I called to ask. Got a job yet?"

Jodi's face flushed. "No. Not even a bite. Tony said Guy Ericson laughed in his face when he suggested me."

Senator Trudeau ignored the bitterness.

"I talked to a good friend about you. He may have a job."

"Who?" Jodi's interest piqued immediately.

"My governor, Jeff Aker."

"Here in Nevada?" Jodi tried to hide her disappointment. She'd hoped the offer would be with Senator Trudeau in Washington. She desperately wanted to pursue the agenda hidden deep within her bosom.

"A great state you'll learn to love," said Senator Trudeau. "Jeff's arriving in Vegas tomorrow. He's staying at the Palace. Get hold of him."

"What kind of job?"

"One right down your alley. He's building a new organization for his next campaign since he barely squeaked through last time. I told him of your success in Nebraska and suggested he give you the top job of organizing his forces."

"Oh, great," said Jodi, concealing the disappointment. But a job was a job.

"The governor jumped at the idea. You'll set up headquarters in Las Vegas. He doesn't want his campaign people underfoot mingling with legislators in Carson City."

"Excellent idea," said Jodi. "Soon office holders want to run the show."

"I've got Governor Aker all primed," said Senator Trudeau in her soft lisping voice. "Maybe I should go into the employment agency business?"

"Senator Trudeau," said Jodi, thinking of her last thirty-seven dollars and groping for words, "I hope to God some day I can ..."

"Forget it," interrupted Senator Trudeau. "For some reason our paths got entwined. You should have no trouble making connections."

"The Governor will think I'm a chorus line, he'll see me so often."

"Got to run. Best of luck," Senator Trudeau said cheerfully before she hung up.

10. DINNER WITH GUY ERICSON

No question Guy Ericson was a handsome devil. Jodi decided that's exactly what he was--a devil in the attire of a blonde Adonis. In contrast to the patient brown eyes of Tony Pellini, Ericson's flashing restless penetrating blue eyes somehow issued a challenge.

"I'm the god Thor. Now who the hell are you?"

Jodi shook her head, she was letting her imagination run away with her. Ericson arrived at the table fifteen minutes late. He spoke in a soft deep voice.

"Sorry. Last minute business," he said.

Jodi noticed the slight smudge of lipstick under his ear.

"Business before pleasure, always," Jodi said sardonically after introductions. She fought the temptation but couldn't resist.

"Hurt your ear?"

Instinctively Ericson reached up with the tip of his forefinger and brought down the evidence.

"No," he said with a grin. "Lipstick. I like lipstick when it's administered in the right way by the right people." He lifted his hand. Immediately a Captain appeared. "Champagne. A good year."

"Yessir, Mr. Ericson." The Captain disappeared.

Guy Ericson coolly appraised Jodi, searching her with his eyes. Jodi somehow felt violated, as if he were probing into her very soul. Ericson, on the other hand, was totally unaware of his prolonged stare. Jodi was not at all what he had expected Tony to produce. Granted, Jodi was beautiful, but she had a strange haunting unfathomable kind of beauty which creeps inside like an invading vine. It could only be grasped up close. Her sensuous magnetism was indescribable. For some reason it annoyed Ericson, as if she exceeded his control. His instincts told him to fight back.

Tony, knowing his boss well, noticed the transformation in attitude.

"See boss," he said enthusiastically. "I told you Jodi had hidden hypnotic beauty."

"Hypnotic." Guy returned his gaze to Tony. "Maybe we can put her in an act...Houdini returns."

"Mr. Ericson," Jodi said with exaggerated sweetness. "I've watched you appraise me like some Nebraska farm animal. I assure you I'm not for sale, so you need have no cost worries."

The statement brought Ericson up short, stifling his next remark. The girl had spunk. A smile flitted briefly at the corners of his mouth.

"I thought you flew out to interview for a job in my legal department?"

"I did. Originally. But when I got here I was told you turned down the idea...cold. So why continue to play games?" She watched the wine steward carefully pour the champagne in the awkward silence which followed.

"I could change my mind," said Ericson.

"Unfortunately, I've already changed mine," said Jodi coldly, trying to hide a quivering lip.

"You really mean it, boss?" Tony jumped into the conversation in an effort to ease the tension. "You'd be picking up a great asset."

Ericson ignored Tony. He faced a girl who challenged his male genes. His spine tingled.

"You think you're qualified to handle legal affairs for gambling, sports, and an international materials conglomerate?"

"No."

Ericson's surprised reaction nearly caused him to spill his drink.

"Then why apply...if you know you're not qualified?"

"A simple carpenter no one ever heard of for 30 years managed in a short time to spread a religion to half the world."

Ericson laughed aloud.

"You think you're another Jesus Christ?"

"Course not," Jodi snapped, refusing to be put down. "But if the world accepted only past performances, new generations would be nothing but pyramid builders for arrogant pharaohs."

"Feisty little gal, aren't you?"

"No more than a Mister Guy Ericson who's evidently been quite spoiled by having his own way."

Guy stared at Jodi for a long moment, then suddenly changed the subject.

"Where'd you stash Miss Zukov?"

"Suite 1101, top floor of the Palace," said a perplexed Tony. He couldn't quite understand Guy's apparent hostility to Jodi. Normally his boss effused grace and charm to beautiful women.

Ericson spoke bluntly, at the same time raising his hand for another bottle of champagne. Tony noticed it was Guy's third glass. Normally he never finished more than one. He sold liquor but thought drunks a bore.

"Accommodations suit you?"

"Perfectly. I've never stayed in such splendor," said a quizzical Jodi. "Why? Planning to kick me out?"

"No. But you've given me the idea," said Ericson. "Just wanted to be sure this pharaoh provided adequate lodgings for a new generation Moses."

"Just let me finish my steak," said Jodi, finally enjoying Guy's obvious annoyance. "Barbs won't injure my feelings. They don't hurt near as much as getting thrown from a wild horse...or trying to survive on an empty belly."

Jodi plunged into the filet mignon and baked potato like it was her last meal, which she was certain it was.

Tony Pellini was visibly upset. In bringing Jodi to his boss he thought the two would be delighted. Their mutual good looks and sharp minds would click together like Cleopatra and Mark Anthony. Instead neither was acting civil. As if a hornet had stung each on the ass the moment they said hello.

Jodi finished her steak, sat back and smiled at Tony. Guy hadn't touched his.

"Umm. Perfect," she said. "As pharaoh you'd make a damn fine cook, Mr. Ericson."

"As pharaoh I'd enjoy driving troublesome malcontents into the Red Sea."

Jodi sensed for a certainty she had enjoyed her last meal on the house. The thirty-seven dollars in her purse shrunk smaller and smaller as her final rampart against hunger.

11. GOVERNOR AKER

At ten o'clock the next morning Jodi called to see if Governor Aker had arrived from Carson City. He had. He was in conference in the Blue Room on the Mezzanine. She left word and decided to take a luxurious bubble bath. It might be her last chance before she got an evacuation call from the front desk. Her dinner with Guy Ericson had been a last full meal on the Titanic. She also resolved to wash her hair.

It was noon when Jodi finished dressing. She thought perhaps the best way to catch a busy Governor was to loiter outside the meeting room and grab him on the run. She picked up her purse and was about to leave when the phone rang.

"Miss Zukov? This is Governor Aker," the voice said.

"Yes. This is Jodi Zukov."

"Senator Trudeau had some mighty nice things to say about you."

"She's a gracious lady."

"That she is...And a damn fine senator, too." The Governor's voice was strong, deep and pleasant. "I'm swamped with meetings until 4:30. Then I'll be at the pool. Need my daily therapy of twenty laps. How about joining me there?"

"Well," Jodi hesitated, wondering what the devil to wear. She'd brought no pool clothes. Only her purple terry cloth bathrobe. A gift from her father, it was so old the elbows were thread bare.

"We can talk and have a drink." He suddenly asked. "Know how to swim?"

"Oh yes," she smiled. That was one of the things you learned early in Nebraska.

"Good. See you in the pool." He hung up.

With no bathing suit? Jodi felt trapped. Thirty-seven dollars was not going to buy a decent bathing suit in Vegas. Moreover, she needed money for food, not clothes. After fretting a few moments she decided there was nothing else to do. She called Tony. Fortunately he was in his office.

"Hold still," Tony said with a chuckle. "I'll be right over."

* * *

Jodi was standing inside the door when he arrived. She grabbed his arm like a lost hiker as they walked down the hall to the elevator. He listened with amusement to her tale of woe.

"You can always charge things to your room?"

"Then Guy would throw me out for sure?"

"I doubt it. You've had a peculiar effect on the boss. He's waging some internal war to determine whether to accept or reject you."

"Hope he doesn't decide until I get work."

"My guess is the odds are in your favor," Tony said. "Still I wouldn't push my luck. He can be cold as steel underneath that handsome boyish face."

Jodi shivered. "He's already confirmed that side of his nature."

"Learn how to lead him. Then he's soft as mush. Pushing him is a mistake, unless you want to tackle a man-eating tiger backed into a corner."

Jodi laughed. "Sounds like my dad. Male chauvinism hiding under a personality disorder. All males seem to have it."

"All? Careful," said Tony. "You could wind up in the pool naked."

"Sorry. Present company excepted." Jodi added with a grin, "I think."

"Want my honest opinion?"

"Not really. But I'm likely to get it anyhow. Right?"

"I think you and the boss are trying to out-gun each other. Two fighter pilots sparring for the kill."

"He seems to bring out the worst in me?"

"The boss thrives living on the edge. Look at his acquisitions. Fast cars, tough professional football teams, gambling, fast horses..."

"Fast women...?"

Tony stopped in his tracks, turning to face Jodi.

"Not so. Though I think he likes to give that impression since his wife died."

"What happened."

"She was running around with another man. They were both killed in a high-speed drunken boating accident in the early morning hours on Lake Mead."

"With a good looking young man at home who had everything? She must have been crazy."

"Or ignored," said Tony. He put up his hand, indicating he didn't want to talk about it.

"I know a shop which specializes in bathing suits for a young beauty with the figure of a goddess," he said. "Come on. Then we'll grab a bite to eat."

It took two hours to select a swim suit. Tony couldn't figure how something weighing only a few ounces could take so much time. But he didn't complain. Watching Jodi try on different bathing suits was like having a ringside seat at a beauty contest. She had an exquisite female figure sculptured to the frame of a raw-boned skeleton. It was as if a sculptor had first built a chassis of classic proportions, then filled in the wide shoulders, firm proud breasts, contoured buttocks and strong chin with pliant clay before bringing it to life. Her nose dropped straight from the line of her eyebrows with only the trace of a bridge to flaring nostrils above full lips. Her mouth contained the only extra flesh on her face.

Tony felt a twinge of bitterness at not being in the race to capture such a prize for himself. Her friendship was all he could ever expect. On several occasions Jodi had subtly made that quite clear.

The swim suit Jodi finally selected was a skin colored nylon-spandex which caused Tony a double-take to see if she was wearing anything at all. It was a two-piece with a top center twist beautifully accentuating the bust. A high-waisted bottom fit snugly around the curve of the hips and allowed for a full display of long shapely legs.

Tony wet his lips. Jodi always seemed to arouse something new. It was as if she had a dozen personalities.

On the way back to the hotel Tony darted inside a drug store to pick up some after-shave lotion. Jodi idly surveyed the traffic and reveled in the warm afternoon sun. She watched three hoodlums across the street swagger their way down the middle of the sidewalk, pushing pedestrians aside. No one complained, accepting it as part of the everyday scene in American cities.

Suddenly, the biggest one grabbed a woman's purse and the three darted across traffic towards Jodi's side of the street. The woman's screams only created idle curiosity. Blasting their way through two women in front of Jodi the hoodlums came straight at her, forcing her to sidestep out of the way.

As the leader started to dodge around Jodi she stuck out her foot, tripping him into the curb. The moment he crashed face down into the concrete she was on his back, jerking his free arm across the curb and landing hard on it with both knees. The howl of pain and loud rifle-like explosion from the crack of a broken arm were simultaneous. The man's arm was shattered below the elbow.

Tony exited from the store just in time to witness the whole scene. He rushed to her side, grabbing the youth by his collar and pulling him erect by the back of his belt. By this time the purse-snatcher was whimpering. The other two youths vanished in the crowd.

Jodi smiled at Tony as she got slowly to her feet, brushing the dirt off her slacks and picking up her packages.

"He's all yours."

Tony returned the purse to the grateful woman and held the howling youth until the arrival of the police.

"Better get him to a hospital. He's got a bad break," Tony said.

A police shake-down produced a switch-blade knife with a six inch stiletto blade. One officer held it aloft.

"You were lucky, young lady. This knife is a killer."

"Not anymore."

Tony and Jodi walked together in silence for a time.

Finally Tony gave Jodi a sidelong glance as he ventured accusingly.

"You enjoyed that?"

"I once roped calves at rodeos. I enjoyed that more."

"I mean breaking the guy's arm?"

Jodi gave Tony a slow steady smile.

"I'm afraid I did. My only regret, I didn't get to break the other arm carrying the purse."

"You plan to break the bones of all criminals?"

"Since we're not allowed to execute mad dogs we can at least break their legs so they can't attack society," Jodi said. "When one man kills another I don't believe in supporting him at $25,000 per year for the rest of his life. Adds to more than one million per lifetime. Some criminals find a secure way of life in our prisons."

* * *

Jodi entered the pool area wearing her purple terrycloth bathrobe which she had worn all through University. She tossed the bathrobe on a chair, stepped out of her sandals and slipped into the deep end of the pool. She hadn't enjoyed the luxury of a swim in a long time. She was on her fifth lap when she looked up to see Tony standing, grinning down at her. Behind him stood Guy Ericson and a big man she immediately took to be Governor Aker.

"An accomplished swimmer too," said an admiring Tony.

Jodi swam over to the exit steps. No use explaining she swam with her dad in the Blue River in Nebraska when she was five years old.

"We've been watching you," said Guy Ericson. "Maybe we should put you in an aquatic ballet. Haven't had one since Esther Williams."

Jodi felt naked before the stares. She grabbed the worn garment and donned it hurriedly as she acknowledged Tony's introduction to Governor Aker.

"Sorry," she said. "Thought I was early. In time for a quick plunge."

"No problem," said Governor Aker, wearing white duck slacks pulled over his swim trunks and a loose tunic. "Meetings were canceled so I got here sooner than expected."

The group settled at a table under a large umbrella, ordering ice tea and cokes. Guy Ericson had a beer.

"Quite an exhibition you put on," said Ericson.

"Hardly an exhibition," said Jodi. "I was enjoying a private swim."

"You create an exhibition wherever you are," said Ericson.

From his tone Jodi couldn't tell whether he was being caustic or complimentary. She turned to Governor Aker.

"Senator Trudeau proved to be one of my better friends. I hope to thank her for this opportunity of meeting you."

"She's one of the finest Senators to come from Nevada. The people love her. Last time around no one was willing to run against her." He paused after a hearty laugh. "Wish I could say the same. Half the state wants to scalp me."

Jodi immediately liked Governor Aker. A hearty outgoing man with the florid complexion of an outdoorsman, Governor Aker's humor started deep in his belly and rumbled upward into his throat.

"I can't imagine why," Jodi said honestly.

"Making decisions alienates people. The tougher the decision the more you build a dam of antipathy."

Guy Ericson rose to his feet.

"Tony and I have work to do. The Governor tells me you two want to talk." With that he and Tony abruptly left.

Governor Aker smiled. "My best friend. A brilliant man. The death of his wife momentarily threw him a curve ball, but he'll snap out of it."

Jodi said nothing. This was no time to express opinions.

"Senator Trudeau convinced me of your abilities," said Governor Aker. "I'm not one for playing games. If you work like you swim, you've got the job as my campaign manager."

Jodi was nonplused. Such directness momentarily pulled the props out from under her. She had expected to detail her work in Nebraska and in Washington and be tested on her legal qualifications.

Taking a deep breath, she said. "I'll be equally direct. I need a job. And I'm certain you'd be a wonderful boss."

"Good. Then it's a deal?"

"A deal," said Jodi.

Governor Aker liked this fresh strong young thing facing him. Though he couldn't quite put his finger on it, he had a gut feeling that she was tough enough to straighten out his bickering operation.

"You've got all the qualifications." Governor Aker eyed Jodi somberly. "A quick sharp mind, experience in campaigning, education, beauty..."

Jodi laughed, gazing about the pool at the lounging women.

"A good looking woman in Vegas is like adding a drop of water to a thunderstorm. It isn't noticed."

"Beauty isn't always the best looking. The most unusual looking beauties are the standouts," the Governor said. "Guy claimed you had an earthy quality. Should help in your relations with people."

"About the job? When do I go to work?" Jodi wanted to change the subject.

"I'll put you on the payroll starting tomorrow." Governor Aker dropped his casual demeanor. "Once aboard I expect a tight ship."

"I'll do my best."

"The job pays $40,000 per year...one third of what a top staffer in Washington makes."

"At least I won't be on a government payroll," Jodi nodded understandingly.

"Right. Guy is furnishing my party workers office space, facilities, telephones, computers and supplies. Headquarters will be located in one of his hotel annexes where he's providing workers with free room and board until after the election. In your case such freebies should nearly double your salary."

"I hate to be beholden to Guy..."

"You're not. I am," the Governor snapped, then relented, adding softly. "We planned this campaign long before you came to Vegas. He didn't have any idea who I would appoint. He doesn't know I selected you. He gave me a list of five worthy candidates."

"Sorry. I wasn't thinking."

Governor Aker spoke cheerfully, ignoring the interruption.

"You'll have the latest technology at your disposal. A real advantage in having a backer like Guy Ericson."

"Workers? Do I solicit my own? Or is there a pool of volunteers to draw from?"

"My staff will furnish a list of names."

"Volunteers?"

"About 15 will be paid, the rest volunteers."

For the next half hour Governor Aker discussed details with Jodi. Finally he offered some advice.

"You can't command troops until you're general. And a general must first be a good soldier."

Jodi shifted her chair so the table umbrella shielded her from the sun. She anticipated a military lecture. She was used to them, even enjoyed them on occasion. Both her father and grandfather had repeatedly enlightened her on the achievements of their distant ancestor, Genghis Khan.

Her favorite was Genghis Khan enjoying a hearty feast of yak and wine spread on a table under which crouched an old enemy. For dessert Genghis had the man dragged out and watched with amusement as aides poured molten silver into the screaming man's eyes and down his throat. A Governor of Mohammed II, Persian ruler, had killed the Khan's ambassador and robbed his supply train of 500 camels loaded with furs and silver. Genghis merely provided the greedy fellow with his fill of silver.

"Political campaigning can be frustrating and tedious," said the Governor, bringing Jodi back to the present. "Politics is probably the toughest skill in the world. Even doting parents can't always please their children. When discipline is a must, temporary resentment flares."

"It certainly demands tough hides," said Jodi, remembering her humiliation at the hands of the Speaker.

"Get out among the people. Talk to them and above all, listen. Give me your feedback. Don't preach. Leave that to the loud mouths. They appeal only to the bigoted few."

"Dad maintained Americans don't wear their ideas as they wear their shirts--a different one each day. Find the constant heart under the shirt. Like the doctor with a stethoscope--listen carefully."

"Right," said Governor Aker, running his hands through his shock of thick reddish hair. "Soon you'll be surprised at the number of faithful supporters?"

"Oh?"

"When they've come to trust you. A condition not gained easily. No matter how painful, keep your word."

"I know," said Jodi. "*Read my lips* destroyed a President who didn't keep his word."

Governor Aker stood up. "I've got to change. Meeting Guy for dinner. Call my office. They'll clue you on all the details."

The Governor hesitated, wondering whether he should warn Jodi.

"Incidentally, I got a call from Speaker Metzer here at the pool just before your arrival."

"Yes," said Jodi, frowning.

"He traced me through my office when he found I was going to speak to you about a job." Jodi could see he was reluctant to proceed. She didn't encourage him, waiting breathlessly for the bad news. Another sacked client.

"He's no friend of yours," he said, finding it difficult to get out the words. "For that matter he's no friend of mine either. Tried to back-stab me in the last election."

"Yes. I guess I should have told you," Jodi said. "Metzer fired me. It had nothing to do with my work."

"I've heard rumors. Most likely lies. I pay no attention to rumors from malevolent foes. Thought you ought to know so you wouldn't lie awake worrying about whether such crap would reach my ears and influence my judgment. It won't."

"Thanks," said Jodi, tears welling in her eyes. She resolved never to short change Governor Aker in her loyalty.

12. A MIDNIGHT CALL

The telephone had an insistent ring. Jodi let it warble its irritating pitch five or six times before struggling awake and turning on the light.

"Yes. Who is it?"

"Guy Ericson," said the voice. "I need to talk to you."

Jodi picked up her wrist watch on the night stand.

"Twelve-fifteen. After midnight. Can't it wait until morning?"

"No way." He sounded distraught and angry. Jodi wondered what insane quirk now possessed him. "Anyhow, it's still early in Vegas. The last show hasn't started yet. Slip on a jacket and slacks. I'll meet you in the lobby." He hung up.

Jodi was furious. The arrogance of the man. Her feelings didn't matter. Yet she was in no position to argue. She was still sleeping and eating free in his hotel. Discretion might be the better part of valor.

Guy Ericson watched Jodi stride, not walk, across the lobby in cowboy boots. He got the impression high heels would not have withstood the impact of her long stride. Only women who took short mincy steps could be comfortable on high heels. Women had to stay on top of the heels or feet could easily slide out from under.

Jodi wore purple shorts and a soft yellow silk bolero tied inches above her belly button. Her bare legs and bare waist glistened with drops of perspiration. The scowl on her face voiced her disapproval.

Guy Ericson took Jodi's arm and guided her past a line into the lounge where an anxious captain hovered over a reserved table.

"Champagne," said Ericson. "My favorite brand."

"Yessir, Mr. Ericson."

Jodi interrupted. "I'll have coffee, please."

"Champagne," Ericson repeated firmly, waving away the captain.

Jodi bit her lip. She felt Ericson's eyes penetrating her body like hot flashes of fire. She felt uncomfortable. For the first time in years her breasts tingled and she felt like a woman. He turned away as she looked into his cold blue eyes. On the ranch the blacksmith taught her cold blue flame was degrees hotter than orange or red. The heat reached below her belly button. She wanted to flee but she sat glued to her chair.

Jodi imagined a blonde Viking was about to grab her hair and drag her onto his ship. They had done that to English women when they raided Britain's shores.

A 10th-century poet wrote of the Viking.

"Blond was his hair, and bright his cheeks. Grim as a snake's were his glowing eyes."

Viking warriors "laid waste to the remains of Charlemagne's Frankish Empire, voyaged to Greenland and were the first to discover America. They colonized Iceland, pillaged Ireland and put France to the fire and sword. They sailed four times to the gates of Constantinople and fearlessly attacked the mighty Byzantine capital. They roamed Russia. Their swift ships, with dragon-headed prows, were the most important technical advance of their time."

William the Conqueror, a Viking descendant, subjugated England and changed it forever into a seafaring nation.

Jodi's instincts cautioned there was something inherently dangerous about this man.

Guy had taken his father's loose coordination of companies and hammered them into a tight ring of steel controlled at the center by Mr. Ericson. His acquisitions had been all or nothing. Only in Material Resources International, Inc., dealing with the defusing of high powered weaponry, did Ericson defer to an outsider...and that was the President of the United States.

Jodi shuddered at the aroma of violence which surrounded Mr. Guy Ericson.

A long silence. The captain returned with the champagne and two glasses. When he finished pouring and left after the usual ritual, Guy spoke harshly.

"You perpetrated a big lie."

Shocked and indignant, Jodi's face flushed with anger. She started to rise from the table.

"I didn't come here to be maligned. I knew better than to respond to your call. Now I'm sorry I did." Her fury nearly brought tears.

Guy Ericson reached over and laid a hand on Jodi's arm.

"You're not Tony's girl at all?"

Jodi slowly resumed her seat. Guy Ericson was incensed about something so trivial she'd completely forgotten about it. She laughed.

"Tony told you?"

"The whole messy story."

"Nothing messy about it. I was on my way. Tony was merely trying to circumvent your bigotry about female lawyers."

"All lawyers, not necessarily female."

Jodi, more relaxed since she surmised the true reason for Guy's behavior, looked at her watch.

"Well past midnight. If you insist on maligning me I may as well get something to eat."

"You're pretty good at eating," Ericson said stiffly. "Some day that habit will ruin the figure I saw displayed at the pool for the ogling tourists."

"Weren't you one of the tourists?"

"I own the place."

"No less an ogler?"

For the first time a smile tweaked the corners of Guy's mouth. He signaled the captain.

"Better feed that undernourished body."

"Good." Jodi ordered a tuna sandwich. "With my mouth full I can't answer your impertinent questions, Mr. Guy Ericson."

Ericson gave Jodi a slow, easy smile. She felt herself falling into a trap. It was time to leave.

"On second thought, food I don't need. I have to get up early. I'm glad we straightened out the Tony misunderstanding."

"You can't leave now. We haven't finished our conversation."

"I know of nothing more to talk about."

"What would you say if I told you I've decided to hire you after all?"

After Jodi's initial surprise, she studied Guy carefully. If only he had a beard he'd look exactly like a Viking raider she once saw in a picture book.

"I'd say no. I'd say I made a mistake in initially misleading you. For that I'm sorry. But I don't intend to repeat bad judgments."

"Legitimate legal work?"

"No. Not now. Not ever. Besides I've got a job."

"With Governor Aker? He told me. A temporary campaign job. Those jobs don't last forever."

"Long enough for me to get my feet on the ground," said Jodi.

"I'll double your pay?"

"No thank you."

Guy Ericson ordered a second bottle of champagne and said little as he sat back and watched Jodi eat. Nothing dainty but ungodly sexy as he felt her mobile lips and mouth devour him along with the sandwich. His body seemed drawn into the magnetism of her earthy aroma. Like a black hole pulling a star into disaster. He tried to shrug away the feeling.

Guy suddenly made a decision. He was going to possess this woman. For the first time in his life he had run into something he couldn't have. It bugged him.

When Jodi finished eating Guy grabbed her hand. The band was playing in the next room.

"Let's dance."

"I'm not very good at it," Jodi protested.

"I am," Guy said.

As they danced Guy pulled Jodi ever tighter to him. At first she stiffened, then gradually melted against his body, becoming one on the dance floor. At times he barely moved in place, entrapping the lithe, full bodied woman encircled in his arms.

Jodi said nothing, feeling his warmth and masculinity and occasionally his male hardness. It had been a long time since University when she loved to dance. Her feared rustiness soon vanished. She wondered at her own behavior. Never before had she allowed a partner to hold her in such a fevered embrace that she felt his manhood. Her conscience stirred uneasily despite her reluctance to break the spell.

13. GOVERNOR AKER'S MANAGER

Although Jodi was responsible for the entire state, Governor Aker placed her headquarters in Las Vegas to keep the politicians in the capital at Carson City from sticking their *cotton-pickin'-fingers* into campaign operations.

"They'd give advice to mating elephants if somehow they could squeeze in the middle," he said.

It took Jodi two weeks to assemble all the Governor's paid workers, who were established as team captains. Jobs were allocated according to geographic, population, industrial, agricultural and other considerations. A brief get-acquainted happy-hour was conducted before the dinner meeting.

Jodi Zukov was introduced to team captains from the podium by Governor Aker, who left immediately.

"The ball's in your court now," he said jovially as he departed.

Jodi faced a group more curious than hostile. She was the only Nevada non-native. They'd heard rumors about her beauty and background. Most were loyal fans who believed Governor

Aker could do no wrong, so they instinctively rooted for Jodi to succeed. In addition, they were being paid.

"Good evening, ladies and gentlemen," said Jodi. "You undoubtedly know more about campaigning in Nevada than I ever will. But as Governor Aker explained, I am here to quarterback the knowledge and efforts of our small team."

"She said our," whispered one delegate to her companion. "That's at least a plus. The last guy called us '*vineyard laborers*' like we'd just crossed the border from Mexico."

"Today," Jodi said, "in consultation with those who have expertise in different areas, we will assign captains to their respective chores. Each captain will be totally responsible. No vague intrusive over-sight committees. Each of us will take total individual responsibility for his or her assignment.

"One captain will handle TV, radio and newsprint. Others will be separately responsible for industry, agriculture, consumer groups, unions, and city, state and federal employees. Captains will report to me twice each week, daily if needed, on what's happening and what improvements are needed. A back-office publicity department will be created for your support. Captains will tell that department what they need. The old policy of the Public Relations people shoving something outlandish down your throat will not be tolerated. Those in the field know their people better."

Jodi brushed back a lock of hair from a forehead beaded with perspiration. Was she getting through, or were they politely sitting on their hands.

"I'm here for support, not to boss. Political campaigns are won or lost on the support of the people. Support, not dictatorship, is needed to win. If I can't get your support, how can you get the support of the volunteers--and they of the voters? It's that simple."

One woman said aloud. "That's a switch. We've always had nothing but bosses."

Jodi smiled. "You captains will be the new bosses. Yet I plan to hold each of you accountable for the success of your assigned group. No finger of blame can be pointed at anyone else."

"Swim or drown, that the ticket?" yelled a laughing young woman wearing a tennis headband.

"Convince the voters to throw you a lifeline," cried Jodi to an applauding audience.

With that meeting Jodi plunged headlong into Governor Aker's campaign. His friend Alex Wallace was running at the same time for reelection to the U.S. Senate for his fourth term. Often their efforts overlapped and Jodi was asked by Governor Aker to assist the older man whenever possible. She did so without question.

Guy Ericson dropped by her office regularly, helping in a hundred small ways. Providing free limousine services to haul state leaders, sending over lunches to harassed workers or clearing the way past bureaucratic hurdles. At the same time Guy was pulled tighter into President Roland's circle of top-secret advisors on dismantling and defusing the highly dangerous weaponry of the former Soviet Union. President Roland awarded MRI additional contracts and placed the company in a top-secret classification to work with the CIA.

The President's foreign policy recognized that the entire Eurasian landmass was ripe for trouble. Iran was getting bolder in its subversive acts of terrorism. President Roland openly condemned the Shiite regime and slapped down all efforts to buy or steal atomic weaponry. Other UN leaders, while agreeing, lacked the courage to step up and publicly label Iran for what it was: a renegade nation. President Roland considered the United Nations an incompetent lottery of unreliable votes.

Jodi's relationship with Guy Ericson drove her into wild dreams which she tried to ignore. On the surface everything was proper. He made no advances, though often taking her to late dinners. When he touched her, even accidentally, she felt a wild, animal sexual vitality assaulting the tingling nerve fibers in her body.

Jodi exulted in the feeling, yet feared his presence, as if he had the power to shatter her defenses. She had never experienced such physical reaction. Frightening to one who had always known she was in control.

With the election only one week away the polls showed Governor Aker headed for a landslide with Senator Alex Wallace winning comfortably. Governor Aker graciously attributed any victory to Jodi Zukov's untiring efforts.

"The gal's a miracle worker," Governor Aker enthused to the group gathered about the pool.

Guy Ericson smiled. "With brains popping out both sides of her head."

Tony Pellini said slyly, tying a new knot in his bathing suit trunks.

"I recall nearly getting canned for bringing a female lawyer to Vegas. Anybody here remember?"

"Not I," laughed Guy, rumpling Tony's hair. "Must have been some dumb cluck...like your old boss."

Guy Ericson's bitter memory of his wife had faded. His renewed interest in business, his involvement with Governor Aker's campaign, his increasingly close ties with President Roland and his intrigue with that unfathomable Jodi Zukov, had all combined to make him forget. He was his old reckless eager self again.

Governor Aker ordered a vodka and tonic from the waiter who hovered attentively over the eminent trio.

"Last time around at this stage of the game, polls showed I was headed for defeat," said the Governor.

"Media damn near called the shots," said Ericson. "You squeaked in by a fingernail."

"Hired a bunch of amateurs," snorted Governor Aker.

"Well," said Tony, barely hiding his gloat. "Some people keep their eyes on the horizons for diamonds, others find them underfoot."

Guy Ericson took the dig in good humor.

"There are miners and there are visionaries," he said, his blue eyes sparkling. "I was looking at the finishing line before I got a horse."

"Got me a winner this time," exclaimed Governor Aker, his looks daring the others to claim the prize. "I hired Jodi while you guys sat on your thumbs."

"Wonder what she'll do in two weeks? When campaign workers fold their tents and everyone goes home?" Tony mused aloud.

"I plan to give her a job in the state house. She proved her ability as an organizer. My staff needs remodeling."

"Oh?" Ericson stiffened in his chair. "Take Jodi to Carson City?"

"That's where the Governor's office is," said Governor Aker.

Ericson's eyes flickered annoyance, even at his good friend. He hadn't anticipated Jodi moving away.

"Will take a good two weeks after the election to clear out the offices," Ericson said. "I want them returned in the shape I loaned them...spic and span."

Governor Aker glanced at Ericson, surprised and hurt.

"Have I ever left a mess on your hands? There's no big hurry. Jodi remains on the campaign payroll until everything's shipshape."

Ericson immediately realized his boo-boo. Tony glanced at his boss questioningly. He knew the problem. His boss didn't want to let Jodi out of his sight.

"Sorry. I was thinking of the normal tenant," said Ericson lamely. "Occasionally I forget those offices are your campaign headquarters."

"No problem," said Governor Aker, returning to his cheery self. "I owe an eternal debt of gratitude to you, Guy. You've backed me at every turn. Perhaps, some day I can return..."

Guy interrupted him with a wave of his hand.

"Forget such talk. I don't perform favors with the expectation of a return. That's only one step short of bribery."

Guy signaled the waiter.

"A bottle of my favorite champagne. We're going to drink to Governor Aker's landslide victory."

14. THE ASSASSIN

Three weeks after the reelection of Governor Aker and Senator Alex Wallace, President Roland paid a hurried visit to Guy Ericson. He wanted inside information direct from the horse's mouth about Iran's rumored purchase of atomic parts. The CIA had given him a lot of dissembling double-talk. President Roland knew Material Resources International operators in Russia didn't trust the CIA but related sensitive facts directly to Guy Ericson.

"I'm ignoring the flabby leadership of the United Nations and plan to tighten the screws down hard on Iran," said an incensed President Roland. "Are the rumors correct?"

Guy Ericson glanced around the pool where they were both sitting in shirt sleeves. Secret service men had cleaned the area for possible bugs and scanned rooftops. A huge beach umbrella practically enclosing both men kept any distant cameras from lip reading. Conditions with the Middle East had soured in the last few months.

"More than correct, Mr. President," said Ericson. "My people tell me Iran is spending money under the table to former

Soviet old-guard hard liners like drunken sailors. They're no longer subtle about it."

"Sophisticated weaponry in the hands of religious fanatics returns us to the national distrust prior to World War I," the President said.

"With today's mullahs operating under 'laws of vengeance,'" said Ericson, "we're back again to a dozen independent nations thirsting for trouble."

"Bosnia proved how destructive ethnic stupidity can be to lives and property."

"With superpower threats gone, Iran's testing muscle."

"Iranian leaders foment fear and violence. Ayatollahs set down the rules and the only opinion that counts in that renegade country says ***kill all infidels***."

President Roland struck the table in exasperation.

"Since the days of Khomeini they've not backed down one inch from their threats to murder infidels."

"Meaning us."

"Today's arms deals would make Bismarck shudder," said Ericson. "My operators report Iran gave $1 billion to the Russian Federation, $5 billion to China, then another $4 billion more to the Russian Federation and $3 billion to North Korea."

"CIA claims Iran plans to flood $12 billion to former Soviet Union hard-liners for engineers and the very atomic weapons we are trying to defuse. That true?"

"True," said Ericson. "Though $12 billion is far too conservative. Buying or stealing atomic weapons can save billions over starting production from scratch."

"Worse, CIA reports Iran spends another billion to support terrorism," said the President. "They want to send the world a message--that's it's unhealthy to confront Iran. By creating instability Iran remains dominant in an uneasy Arab world."

"Mullahs preach scorn for the U.S. and the United Nations. They declare Desert Storm the last gasp of a dying democracy. Tepid efforts in Bosnia reaffirmed their belief our military is nothing but a hollow loud drum."

President Roland's tired eyes blinked in frustration.

"Americans don't understand the Shiite Iranian mind anymore today than the Jews understood Hitler when he took his first small steps to power."

"Not that we haven't been warned often enough," said Ericson. "Until a Pearl Harbor happens, nothing sinks in."

"Lowering the boom on sales to Iran raised a lot of hell," said the President. "I had to do it. Our Western culture is at stake."

"The Iranian President thinks he's a modern Genghis Khan."

"Iran's Persian." President Roland's tightened jaw muscles expressed his disdain. "Genghis Khan was a Mongol. Mongols beat the hell out of the Persians every time they met, mostly against odds 5-to-1 or better."

"So did Alexander the Great," prompted Guy.

"I know," said the President. "Persians were always better at intrigue than fighting."

"One thing we'd best be prepared for in putting the screws to Iran, Mr. President, is another Hezbollah or Hamas attack."

"Like the World Trade Center? I've spent sleepless nights with that thought. But when a nation grovels before blackmail it ceases to be a sovereign nation."

"When will you cut off all trade with Iran?"

The President smiled. "Minute I return. Wanted to use you as a sounding board. And to get away from Washington. Any objections?"

"No. Would my comments make any difference?"

"Probably not," laughed the President. "But I do value your opinion."

"I concur 100 percent. I'll warn my operatives to be doubly watchful for sabotage," said Ericson.

The President looked at his wrist watch.

"Now to other matters. I asked Governor Aker to join us. His landslide victory puts him in a class by himself."

Guy Ericson stood up and peered beyond the secret service operatives scattered unobtrusively in front of the glass walls. He spotted Governor Aker sitting inside and waved.

"He's waiting for us to finish."

President Roland greeted Governor Aker cordially.

"I came out here to take some lessons," he laughed. "All I hear are glowing reports on how you managed your campaign."

"Jodi Zukov deserves the credit, Mr. President," Governor Aker said with a broad grin.

The President turned to Ericson.

"I've got to meet this genius. On the phone I thought you two were about to erect a statue?"

Governor Aker emitted an abashed smile.

"In the euphoria of a landslide perhaps I got carried away. Initially I expected defeat. I certainly wasn't prepared for what happened, although Jodi insisted I had no worries."

"Pregame jitters," nodded the President, understanding.

"As an objective outsider, the returns fooled me too," said Ericson, struggling his fingers through tangled blonde hair.

"So tell me, what'd the gal do that was so unusual?"

The Governor accepted an ice tea from a waiter while gathering his thoughts.

"Jodi was a brutal organizer," he said at last to the President. "She weeded out non-essential topics and ruthlessly prevented the media from pursuing personality foibles. She researched the basics and addressed only those subjects which concerned the people."

"How the devil did she control the media?" A deep frown of annoyance creased the President's face. "Most got the persistence of barnyard mules."

Governor Aker said. "They drove me daffy too, until I watched Jodi handle them."

"Her system takes guts," said Ericson.

"And creates media enemies," added Governor Aker. "Yet it works."

"How?" President Roland was intrigued. The media was a festering wound which never healed.

"Jodi stuck to the primary, relevant issues," said the Governor. "When a member of the media raised personal, nasty or irrelevant questions, Jodi would point to the next questioner as if that person didn't exist...or she hadn't heard. Also, she never called on the nasty one again."

Ericson chuckled at the recollection.

"One belligerent sport hollered when Jodi ignored his question. 'Hey, you never answered my question.'

"'Nor do I intend to,' she retorted without missing a beat."

The President shook his head dubiously.

"The White House Press Corps is a collection of hyenas. If they smell blood, the pack surges into a barbaric frenzy."

"When someone tried to draw blood with a back-stabbing quote," Governor Aker said, "Jodi immediately stopped the questioning and demanded sources. If the questioner couldn't produce, she shattered his image before associates by implying that possibly the questioner was an out and out liar. She dismissed anonymous sources as the garbage of malicious minds and suggested the questioner go back to Journalism 101."

"Sounds brutal? Surprised she got away with it?"

"She got ultimate respect, Mr. President," said the Governor. "Most ended up carrying her torch. She made their work easier by sticking with facts and pin-pointing the state's problems. She presented my platform in clear unmistakable terms. She made me a hero instead of a politician."

President Roland leaned back, relaxed. The weather was ideal and soft breezes drifted over the pool to freshen the air.

"Sounds like you found another silver lode here in Nevada. I could use Jodi in Washington," the President said slyly.

"In a pig's eye, Mr. President," Ericson exclaimed without thinking. Softening his tone, he said. "We need her in Nevada. Lot of work to be done here. Besides, she came from Washington. Speaker Metzer fired her."

The President's face colored.

"The Speaker's a pompous ass. Got the brain and looks of a dinosaur. Big rear end supporting a pea brain on a fat neck."

Governor Aker spotted Jodi Zukov through the plate glass. She was sitting inside having iced tea with Tony Pellini.

"I see Jodi in the lounge, Mr. President. Would you care to meet her?"

"I'd be delighted," said an amused President, for Governor Aker was already on his way to fetch his star.

When Jodi and Tony were introduced, President Roland glanced at Governor Aker with a quizzical smile.

"You describe an Aristotle and introduce a Cleopatra. Small wonder you're trying to hide her in Nevada." He turned to Jodi. "They tell me you opened their eyes to new style campaign techniques."

Jodi blushed. "Eye witnesses are not always reliable, Mr. President.

Jodi wore a soft purple blouse trimmed in gold with a matching purple headband securing her hair until it dropped evenly across her shoulders. A tight purple skirt with a braided golden belt completed the simple ensemble.

"Beauty is Jodi's camouflage," teased the Governor.

"Hardly a good one," declared Ericson. "A heat-seeking missile would find her hot temper every time."

An embarrassing silence hung in the air after Ericson's sudden eruption. Tony recognized the barbed remark as a frustrated slip of the tongue. The others, puzzled, tried to ignore the outburst.

A burst of rifle fire shattered the peaceful scene and echoed back up into the canyon. Bullets splattered against concrete and ricocheted across the patio, smashing plate glass windows. Secret service operatives rushed to the President, throwing him down and piling their bodies on top of him. Men with telescopic rifles peered into the nearby foothills from whence the shots came. Walkie talkies alerted highway patrols and guided the search. Pandemonium forced terrified guests to scramble indoors and distance themselves from the President's party. The shooting stopped as suddenly as it began.

Initial concern was for the President. Secret service agents pulled a shaken but grimly determined President to his feet.

"I'm okay," he insisted, helping an operative return the beach umbrella to upright. "How about the rest of you?"

"Fine," said Governor Aker.

"Me too," echoed Tony Pellini, brushing himself off.

Guy Ericson didn't answer. He was bent over Jodi where he had covered her with his body. He looked down at his shirt front, soaked with blood. Still on his knees, he stared at Jodi. She didn't move. Blood was oozing from the side of her head.

"Get the house doctor," Ericson yelled frantically at Tony. "Jodi's been hit."

"Oh my God," cried Governor Aker, staring at the blood spreading under Jodi's unconscious form. "The bastards murdered my angel."

15. THE HUNT

Within minutes the attempt to assassinate the President was broadcast on all airways--police, radio and TV. A dragnet was thrown around Las Vegas and the airports, public and private, were temporarily shut down. After six futile hours of search the bans were lifted, although tight security was maintained at the airports. The President departed for Washington on Air Force One shortly after the attempt on his life.

It wasn't until 48 hours later that a battered old convertible with an outdated license plate was ordered to stop on the strip. The officer merely wanted to question why the occupant was driving on an expired plate. When the officer got out of his squad car the convertible took off at high speed, belching smoke.

After a wild chase, with another police car blocking his exit, the driver jumped out of his car and began firing back at the police as he headed between two buildings. A fusillade of bullets brought down the fleeing culprit.

From papers on the body the police learned he was an Iranian Shiite Muslim by the name of Abdul Hamil. He had entered the country on a student visa and was working in the

Palace Casino as an assistant dishwasher. He had used a cheap automatic hand gun to fire at the police. In the trunk of his car they found an expensive telescopic rifle used in the assassination attempt. Bullets from the rifle matched those retrieved from the pool area and from inside the shattered glass wall of the lounge.

"Fools luck," grumbled one FBI agent in discussing Hamil's death.

FBI agents and the local police were assembled in the property room of the Las Vegas police department examining the rifle.

"Mighty damn fine weapon," said the property officer, turning the rifle over in his hands. He wore surgical latex gloves. All finger prints and muzzle firing requirements had been performed. He was in the process of turning the weapon over to the FBI for transfer to Washington.

"No dishwashing bum can afford a weapon like this," said agent Roth. "Must have cost at least $3000 or more. Plus the latest most sophisticated telescopic sights.

"Fortunately, he got the rifle without any training," said a policeman. "Only a stupid amateur could have missed his target with this weapon."

"We've got to find where he got it," said agent MacDonald, frowning. "He couldn't have transported it by air."

"He entered Immigration through New York," said Roth. "We know that. So he must have driven that jalopy all the way out here."

"Hamil's organization obviously knew the President often visited Guy Ericson. Assassinations are easier performed out among the people."

"There's big money supporting the intelligence and weapons network behind their sacrificial lamb, Abdul Hamil," said Roth. "A simpleton like Hamil couldn't execute such a plot."

"When I talked to headquarters again less than half-hour ago," said MacDonald, "they're positive Hamil is from the same cell which bombed the World Trade Center. That was done by amateurs too."

"Trouble is, someone's supplying these expendable fanatics with money. Weapons like this don't come cheap," said the property policeman.

"Evidently not as cheap as the lives of incompetent zealots," chuckled Roth.

"Professionals are usually branded. They're known to us and Interpol," said MacDonald. "Amateurs cross borders with impunity."

"Very reason they're so damn dangerous," swore Roth. "Like matches, light one, it gets blown out; so light another. Who cares how many matches it takes to light a fire when you've got an unlimited supply? Certainly not the mullahs in Iran."

"Right." added MacDonald. "Ayatollah Khomeini didn't lose one night's sleep over losing a million soldiers in the war with Iraq...all expendable matches."

"I took part in the World Trade Center arrests," said Roth. "Terrorists look so feeble when picked up by the police."

"'Cause they're carrying feeble intellects," said MacDonald, his reddish complexion flushed with indignation. "Devoted to blind fanaticism. The more feeble the intellect the more it can be cult controlled to perform incredibly evil behavior."

"Like calm cannibal chain killers in the court room," said Roth, "they elicit feelings of sympathy from elements of our society. But inside their twisted minds lurks an appetite for blood."

"My parents came from Scotland," said MacDonald. "Sheep dogs who once kill and taste blood had to be immediately destroyed."

"Why? They always kill again?"

"Mostly. Primordial instincts return, I guess," said MacDonald. "Or perhaps, like opium, the sensation clicks on receptors previously dormant. The more they're awakened the more addicted and frenzied they become."

FBI agent Roth watched the expensive rifle being carefully wrapped. His mind was on the World Trade Center.

"How can a blind cleric, supposedly of God, harbor such hatred?"

"Like a programmed cult child he has a limited horizon." MacDonald felt an intense sense of frustration. "Hitler's youth, Khomeini's youth, the youth in any cult are rarely allowed information from the outside. They're taught outsiders are the enemy."

Roth handed the wrapped rifle back to the property custodian.

"Treat this carefully. Though they got your ballistic information and fingerprints, our labs can find a clue in an ocean."

"Leaving today?"

"Booked on an afternoon flight. Our superiors want the weapon."

MacDonald explained. "President Roland is furious. He knows Iran is behind all this treachery but he can't put his finger on the exact source."

"He's in the middle of discussions on further actions against Iran now," said Roth. "This assassination attempt will double his resolve to punish their arrogant deceit."

"Why the hurry for the rifle?" The custodian was puzzled. "They got all the information."

Agent Roth winked at the property custodian and shook his head as if the poor fellow did not understand the real world.

"President needs it for display at the White House press conference tomorrow noon."

16. GUY SURPRISES

Jodi's head throbbed unmercifully. Her eyes remained tightly closed. Forcing them open would be painful. She started to lift her hand to her forehead. It wouldn't budge. In near panic at the thought of being paralyzed she squinted tightly, then let her eyelids slowly lift at the recoil.

Glancing down she sighed audibly with relief. A large hand covered her own at the side of the bed.. Twisting slightly she saw Guy Ericson's broad grin come into view.

"Time you woke up," he said brusquely. "Don't think I can hold out much longer. My hand's numb."

"So's mine," said Jodi, trying to flex her fingers under Guy's grip.

"You gave us quite a scare, getting in the way of that gunfire."

"Oh, sure. I jumped out in front," Jodi grimaced.

The concern in Ericson's eyes belied his jovial attempt.

Jodi glanced up at the ceiling. She lay in a large canopied bed. Definitely not in a hospital.

"Where am I?" Jodi felt an instant sense of alarm.

"In my bed," Guy said with obvious amusement, "where you've slept for the past 24 hours as if you owned it. I've had to sleep on the couch." He pointed to the large golden sofa along the wall of the room. A pillow and disheveled blankets lay on the floor.

Jodi's headache was suddenly gone. She wanted to know the details. She remembered sitting with the President and his party when the spray of bullets hit. Then everything went blank.

"What happened? Was the President killed?"

"No. You were the only one hit. A bullet ricocheted off the tile and dealt a glancing blow to the side of your head. The doctor says you suffered a temporary concussion. You'll be fine in a few days."

"Guess I was lucky," said Jodi, gingerly fingering the bandage.

"Lucky the assassin missed, or lucky you've got a hard head?"

"Both. I guess." Jodi looked at Ericson in light blue pajamas then down at herself under the sheet. She was stark naked.

"What the devil?" she sputtered. "Who undressed me?"

"Dr. Stockdale and a nurse," said Ericson hastily. "You were covered with blood. So was I. Dr. Stockdale and I brought you upstairs to my suite rather than move you four or five times in getting you into an ambulance and into a hospital. Dr. Stockdale said with a concussion you should be quiet as possible."

"You brought me here?"

"Yes. Dr. Stockdale and I."

"Now you're in pajamas?"

"Like I said. I was covered with blood too. While the nurse bathed you, I took a shower and slipped into these." He glanced at Jodi with a wry smile. "You don't like the color?"

Jodi's worried frown changed into a broad smile.

"I'm not ungrateful. Just curious. I like to know what happens to my body when my mind isn't keeping track of it."

"You got shot, causing a temporary concussion. Not serious."

"Not serious? I damn near get killed and you call that *not serious?"*

Ericson said. "No worse than professional quarterbacks experience every Sunday."

"I'm no quarterback," said Jodi, incensed.

Ericson smiled. "The doctor gave you some quieting shots last night. The nurse just left. Nothing else happened to your body," he said, adding with a twisted smirk. "Not that I didn't think of crawling into bed with you. It had to beat sleeping on the couch."

Lying nude under a sheet and watching Guy Ericson at close range in pajamas exposed unwanted basic feelings in Jodi. Her breasts tightened and she felt a drying inside her mouth as she looked at his insolent yet concerned attitude. Windswept blonde hair sat atop blue eyes which never left her face. His gaze was direct and disconcerting.

Guy stood up, leaning over her. She looked up and knew something dangerous was about to happen but she made no move to prevent it.

"I like you best when you're quiet," he said. "I know only one way to accomplish that."

He closed her mouth with a kiss, gentle at first then increasing the pressure until her lips were hard against her teeth. Initially Jodi lay unresponsive, fighting the demons surging up within her. When she could stand it no longer she threw caution to the winds and wrapped her arms around Guy's neck, probing the inside of his mouth with her feverish tongue. The barrier sheet dropped away as he drew back to admire the breasts which stood out like marble edifices. He took one in his hand and kissed the nipple. When he finished Jodi pulled his face down against the other.

Guy Ericson made love to her until she could no longer withhold the screams. He gentled her and pounded her until his touch burned fires of desire deep into her belly. Jodi had not wanted this to happen. The cold-blooded Viking bastard. She knew nothing would come of it except to make her his slave. For the first time she understood how women of history became entrapped by domineering men.

He possessed her entire body, like a reptile swallowing a bird. Her breasts, belly button, calves of her legs, shoulders and back, were all enslaved to his touch.

When it was over they lay quiet for a full ten minutes.

His first words were another total surprise.

"Jodi, I love you. I've loved you from the moment I saw you. But I've fought the idea of falling in love like a Roman gladiator."

"Viking," corrected Jodi softly, "like a seagoing Viking, leaving weeping wives and children behind."

"A Mongol should understand," said Guy. "They left behind more weeping wives and children than a plague."

"Vikings robbed, raped and pillaged English coasts like swarming locusts," responded Jodi. "And with no more morals."

"Spare me a Mongol lecture. It's like advice from a hangman."

Jodi snuggled into Guy's arms, tracing a forefinger across his chest.

"Perhaps we deserve each other?"

"Mixing Mongol blood with Viking blood has the portends of a Shakespearean tragedy."

"Cowards die many deaths, the brave but one," Jodi quoted with a light laugh.

Guy turned abruptly on his side and pulled Jodi tightly against him.

"Marry me, Jodi," he said. "I'll take you one more step from your Mongol beginning."

The sudden unexpectedness of Guy's proposal shocked Jodi. She bit her lower lip hard. What was she getting into?

"You've got sex confused with love?"

"No. I've really known it from the beginning...just refused to face it."

"Maybe we both should think it over. My dad always claimed major decisions should be made either on horseback or on your feet."

Guy jumped erect alongside the bed. "Now I'm on my feet. Marry me."

Jodi laughed aloud at the image of Guy Ericson in the nude standing at attention.

"A Roman sculptor would delight in you as a model."

Guy immediately climbed under the sheets, straddling Jodi's body and peering straight into her eyes.

"Men usually propose on their knees. Only a Mongol would want them on their feet. So what's your answer?"

"God plays mysterious tricks. Maybe..." Jodi's voice trailed off.

"That a yes or no?"

"It's a maybe. I like riding bucking broncos occasionally at rodeos, but I've no desire to spend my life on top of one."

"You like making love to me? Now be honest."

"Oh yes," said Jodi, running her hands up the smooth sides of his athletically honed body. "I love it. And I'm certain I love you. But there has to be other things to do during the day."

Guy kissed her on the mouth and the breasts, firing her body again with his powerful hidden current.

"Let's try another siege of love making. This time I want you to concentrate on what's it's like to have me coming home every night."

Three weeks later Jodi Zukov changed her last name to Ericson. They were married in a private ceremony in Carson City with Tony Pellini as the best man and Governor Aker giving the bride away.

Before formal rites, however, an understanding was reached between Jodi and Guy.

"Since you didn't want me in the first place I will not work for you," Jodi told her prospective husband. "Governor Aker wants and needs me."

"I need a good lawyer too...One to reign over my legal eagles."

"I've thought it over," said Jodi. "We're independent mavericks. Our marriage would never last if we both tried to sit the same saddle."

Jodi smiled at Guy's crestfallen features.

"You'll be happier without me underfoot. Governor Aker is creating a position, *Consultant at Large,* responsible to him personally. He will establish my permanent office in Las Vegas, so I'll be at home."

"*Consultant at Large?* What kind of position is that? A personal valet?"

"No. A job removed from the influence of any office holders or legislators. He wants to remodel the state government in line with all the campaign promises. And he doesn't want to subject my views to a stamp of approval from some bureaucrat."

Guy protested vehemently at first.

"Politicians are despicable weasels...why join such ranks?"

"You highly regard President Roland and Governor Aker?"

Guy fumbled for an answer. "They're friends...exceptions."

"True. In any business you've got to know the white hats from the sleaze. In your operations you've got more security police than most cities. You don't permit sleaze. I expect to exercise the same discrimination."

"I just can't believe you're that interested in politics," fretted an upset Guy Ericson.

Jodi kissed him, capturing his lower lip between her soft plush ones and tugging it playfully.

"Like a rodeo animal, I've been trained for this event and I want to stay the course."

"What the hell you talking about?"

"My father and grandfather were badly abused by bureaucrats in the political system. I vowed revenge. I struggled through law school, performed the meanest political chores, pestered a Nebraska Congressman to get me a job on the Speaker's staff, endured an attempted rape and starved for a year in Washington to obtain some entree into a lousy system that needs correction. I don't intend to stop now."

Guy Ericson stared at Jodi in astonishment.

"That's a lot of baggage for a young woman to carry."

Jodi smiled. "I can handle it. It's not made me stoop shouldered yet. I'll not burden you with it."

"Good," said Guy, a bit mystified. "Perhaps along the way I can help, but I've no intention of getting caught up in some vendetta."

"It's no vendetta. It's warfare for reform."

"Well," shrugged Guy, "there's no question the government needs reform."

The subject wasn't mentioned between them again.

17. THE INSATIABLE MEDIA

Once the media learned of Jodi's presence at the assassination attempt on President Roland, they became insatiable. Why the only woman in the gathering? And why the only one wounded? When they also discovered a cover girl married to one of the nation's richest men, they were in "hog heaven" as Governor Aker scathingly remarked.

"You've become another Jackie Onassis," he quipped. "Photographers will jump at you out of dark alleys and peer at you from ladder tops."

"No skinny dipping," grinned Tony Pellini, "or you'll wind up on English tabloids competing with the royal family."

Jodi winced at the thought. "Thanks...thanks for alerting me to all the dangers. You suggesting I crawl in a hole and hibernate for a year?"

Tony protested. "Not at all. Perhaps you can turn all the attention to your advantage. Like Hearst said, 'call me anything, just spell my name right.' You twisted the media around your finger during the Governor's campaign. I'm sure you can do it again."

Governor Aker declared enthusiastically.

"You better believe it. Jodi's my personal right arm. I'm backing her 100 percent."

It was early fall and Jodi had been in her job nine months, balancing a happy marriage with touring the state to listen to the people. Her final report was ready. There were no surprises to Governor Aker. She had worked closely with him in digging out the problems. But there would be many unpleasant surprises for legislators and bureaucrats. Since it was the Governor's second and final term, reelection held no threats and he was determined to drastically change the old-boy legislative system.

Jodi worked hard, often slaving into the night on a computer until Guy got up and physically hauled her to bed.

"You're not married to Governor Aker," he said in a moment of pique. "You're married to me."

"Yes...I know. And it's you I love," Jodi said, repentant. Guy's touch still sent shivers of electricity through her body. When in bed with him she was in another world.

Jodi walked into her office one afternoon after having lunch with her husband to find Governor Aker in her chair poring over her report.

"I've read this thing three times," he said, waving aloft the fifty-page document. "Now I've come down to hear you present it verbally in two pages."

"A compressed disk or oral shorthand?" Jodi commented after the Governor had given her a big hug and moved to sit with her on the sofa.

"People today want things in sound bytes. You know that."

"The report backs up those quick summaries," Jodi said. "I don't expect the people, bureaucrats or even most of the legislators will ever read it. But you and I need to know what's wrong...and in detail. Because the cures come in specifics. It does no good to treat a strep throat with cough drops."

The Governor studied the intense face of the young woman beside him. She was thumbing through the pages. Jodi was an unconscious beauty, without an ounce of vanity. Her long black hair dropped in waves over her shoulder as she perused the document. When she raised her eyes to his, her smile took his breath away. An exquisite aroma of delicate perfume wafted into his nostrils.

Governor Aker smiled his satisfaction. He had truly been the one to discover Jodi. Now she had become as dear as a daughter, married to a friend who had nearly capsized at sea until Jodi came along. Today Guy Ericson was again a solid businessman. Respected throughout the country. Before he met Jodi he had been on the verge of becoming another irrational Howard Hughes--a man who had everything and threw it away.

"What have we learned?" Governor Aker asked in a quiet tone.

"That Nevada is a microcosm of the national picture."

"No surprise there."

"Legislators, bureaucrats--who number in the thousands, the media and an untold number of agencies make their living by bashing the people who feed them...they just don't get it."

"In my youth I respected the media, almost idolized them," said Governor Aker, stroking his chin thoughtfully. "This past decade I've secretly come to consider them no better than the loud mouth publicity hounds they cater to."

"No question. Those with the megaphones or picket signs get the media attention. Avoids research."

"If scientists researched disease like the media scrutinizes anonymous sources," said the Governor, "we'd still have plagues wiping out nations."

"For starters, Governor, I've unearthed a tax related industry--including city, county and state, spilling over into federal--which extracts more from the citizen than the combined revenues of the steel, auto and aircraft industries."

"Bureaucrats?"

"Without doubt bureaucrats also gouge the public pocketbook, but I'm talking about the people our legislators created with their laws and regulations."

"Who?"

"An industry of special causes, professional lawyers and lobbyists who get high pay for not producing one single ear of corn."

Governor Aker nodded unhappily.

"We keep overlooking the obvious. Yet the media raises hell about agricultural subsidies."

"Food stamps, welfare payments, litigation explosions, energy and environmental regulations and a hundred other dubious programs soak up more money that ever goes into the support of those who produce food," said Jodi curtly.

Said Governor Aker. "We're fast becoming a nation of one-man, one-vote non-producing free-loading constituents."

"A nation of Ponzi operators?"

"Money spent on poverty programs could purchase all the U.S. farmland plus all the assets of the Fortune 500 companies."

"Consider another drain of unproductive money," said Jodi. "There are more private security guards in the United States than public police officers."

"A condition our forefathers never dreamed could happen."

"Nor did they dream congress would become so rapacious that it would legalize plunder."

Governor Aker stood up to stretch his legs and wander over to the window.

"What are you talking about, Jodi?"

"Here in Nevada we're getting a prime invasion of government operating under the guise of '*drug fighters*' confiscating property. Doesn't matter whether the owner is convicted or even charged with a drug crime. Agents get to keep their spoils. Naturally they select the best cars or boats or whatever. So far the government has stolen from landlords over $1 billion in properties."

"Yes, I know about their illegal taking," said Governor Aker, running a distressed hand through his hair. "But those are federal operatives. In the state we're practically helpless."

"They love to operate in the glint and glitter," said Jodi. "Boats, airplanes, fancy cars are their meat, not drug enforcement."

"The entire costly drug enforcement program is a joke," exclaimed Governor Aker. "Soaks up money and accomplishes not one damn thing."

Jodi smiled bitterly. "Like Pascal's 'the mountain was in labor and out came a mouse'."

"The 'taking clause' reads 'nor shall private property be taken for public use without just compensation.'" said Governor Aker. "It forbids petty larceny as well as grand theft. It does not distinguish between taking autos or small boats and large properties."

"A thief cares little about distinctions," smiled Jodi. "Doesn't seem to matter whether the thief is an agent of the government or a common criminal."

"While the Japanese are hiring engineers," said the Governor, "states, counties and the federal government are hiring lawyers. We've now got 70 percent of the world's lawyers."

"My husband says liability costs have no ceiling. Like congressional spending, they go up and up."

"In Nevada the immigration mess is getting to be priority number one in my office. We can no longer afford aliens--good or bad. They destroy our schools, wreck our aid programs and break our financial backs with social programs we can't provide."

"Trouble is," said Jodi reflectively, "passage of Proposition 187 in California aroused an opposition cult of those who pay for nothing."

"And are supported by the courts."

"With judges who never met a payroll."

Jodi's intercom buzzed. Her secretary was apologetic for the interruption.

"Senator May Trudeau's on the phone. Wants to speak with Governor Aker. Says it's urgent."

"Put her through. Senator Trudeau on the line," she said, handing the telephone to Governor Aker.

The Governor listened carefully, his brow creasing into a deep frown.

"Of course, May. I'll take care of everything at this end. We're all terribly sorry. See you soon." He hung up, absently standing idle by the telephone a full minute before turning to Jodi.

"Senator Alex Wallace's pacemaker couldn't handle the job," said Governor Aker, his eyes moist. "Gave Nevada 30 years service. Died this morning."

Jodi was silent. She hadn't known Senator Wallace well except that she'd helped him incidentally in his campaign on the same ticket with Governor Aker.

"Senator Trudeau's coming home with the body."

In deep thought he walked over and sat down heavily in Jodi's chair. "There's a lot to do."

"Can I help?"

Governor Aker looked up, as if seeing Jodi for the first time since the call.

"Jodi...Jodi," he repeated, coming out of his trance. "We hate to see loved ones die."

"I know, Governor. I know," said Jodi softly, thinking of her own family and the bitter struggle which brought on her father's fatal heart attack.

Rousing himself, the Governor straightened in the chair. He looked tired. Small lines had accumulated around his eyes.

"Guy and the Senator were good friends of long standing. In a small state like Nevada there's precious little middle ground. Either stand up and be counted or you're an enemy."

Jodi sat quietly, realizing there was nothing she could do to ease the loss of a long-time friend.

"I really don't feel up to it. Need a fresh input. Definitely no politician."

Jodi was puzzled at the Governor's ramblings.

"Up to what, sir?"

"Up to giving Senator Wallace's eulogy. Your husband should give it." He got up abruptly and went over to Jodi. "Will you ask Guy to give the eulogy? I don't like the idea of either Senator Trudeau or myself taking the spotlight at the funeral. We're both politicians. Critics will call us insensitive and self-serving."

"How totally unfair," said Jodi. "I'm sure Guy will be happy to do what you and May want."

"Senator Trudeau thinks the same way," said the Governor.

Jodi had not seen him so unhappy since the assassination attempt on the President. She knew that invasion of security and her wounding in Nevada made the Governor feel he's personally slipped on security.

18. THE FUNERAL

Jodi gazed into the coffin and thought of her own father. All the work, agony and heart-rending effort to carve out a niche for yourself and your family whisked away in a moment. The gods were remorseless. They didn't seem to care who was right or wrong. Until man stepped into the picture, nature at least had been even-handed about death. Natural deaths had been the norm. Now society prolonged life in more tortuous ways than jackals ripped open their prey.

Senator Alex Wallace's serene face indicated his end had not been difficult. Not like her father's who had collapsed from the stress of fighting off federal bureaucrats who were determined to evict him from his land. Her face flushed with anger as she returned to her seat.

An old mission church on the edge of town bulged with notables. President Roland flew out for the day. He was traveling to Seattle later for a meeting with leaders of the Pacific Rim nations. Senator May Trudeau sat with General Pete Hardin, Air Force Chief of Staff. Across the aisle Speaker Metzer and Senator Mordecai Meyer formed a group of cronies who arrived in Air

Force planes. They planned to return to Washington after first tasting a bit of night life along the strip.

Guy Ericson climbed the steps to the podium.

After a few introductory remarks, thanking the President and those in attendance, he spoke of Senator Wallace's long years of service to Nevada.

"When Senator Wallace arrived in this wild territory his name was Alexei Karopurtkin, a boy escaped from an iron-fisted Soviet Union. He found nothing but sand, desert and dust. But he had a vision. He wanted desperately to be an American. So he became Alex Wallace.

"He fought to keep the individual free. Russia taught him no individual can be free if shackled by taxes and regulations. 'Laws and taxes are chains, however well intended,' was his favorite theme. We would all do well to listen to his words and not wind up like a collapsed Soviet Union."

Guy Ericson recited many of Senator Wallace's accomplishments and closed with the hope his successor would represent the people of Nevada half so well.

That evening, before he left, President Roland sat informally in the Palace Lounge with Speaker Metzer, Senators Mordecai Meyer and May Trudeau, Governor Jeff Aker and Tony Pellini. Guy Ericson and Jodi were on the dance floor where they had been most of the evening, clinging to each other like drowning sailors.

Governor Aker glanced wistfully at the couple.

"To be young and in love. I can hardly remember."

President Roland glanced at his wrist watch. His countenance was stern.

"I've got to run in a few minutes. But I'd like to feel you out on a replacement? I've got some delicate votes coming up in the senate. Or am I pushing too hard, Governor?"

"Not at all, Mr. President. I've about made up my mind. How about Guy Ericson?"

"Excellent," exclaimed the President, obviously delighted. He turned to Speaker Metzer and Senator Meyer. "You two would support such a choice."

"He's not our cup of tea," grumbled the Speaker. "But fighting him would probably do our cause more harm than good. He's got the charisma of a TV god and the power of Thor."

"It's my choice anyhow," Governor Aker reminded the two.

"Fine," said the President, standing up and placing a hand on the Governor's shoulder. "Give my best to the love birds. I wonder how Jodi will like being back in Washington?"

The Speaker grimaced at the remark. After the President left he and Mordecai wandered off to try their hands at the gaming tables.

Governor Aker and Tony Pellini were alone. Tony ordered champagne.

"Calls for a celebration, Governor. You and Guy have been like Damon and Pythias down through the years."

When the waiter appeared and completed the normal champagne ritual, Governor Aker lifted his glass.

"I've rarely been so self-satisfied. Guy becomes senator and married to a gem discovered and nurtured by the two of us."

Tony Pellini experienced a sharp pain in his eyes as he fondly remembered his close liaison and shared intimacies with Jodi when she was a struggling novitiate in Washington. Jodi insensibly had become his ward. She was an unfathomable rascal, capable of unpredictable extremes.

Neither saw the storm clouds on the horizon.

* * *

"U.S. Senator? Romp with that swarm of locusts gone amok. I should say not," Guy Ericson stormed when approached about the job.

Governor Aker's face flushed red with embarrassment.

"The President thinks..."

"Never mind what Roland thinks," raged Ericson. Jodi had never seen him so angry. "Congress is a sick animal which I would destroy if it were a horse."

Governor Aker bit his lip, totally overwhelmed at his friend's response to what he thought was a princely offer.

"I'm sorry," he said stiffly. "I'd no idea you'd react this way. I should have consulted you before I mentioned it to the President."

"Right," said Ericson, running both hands through his hair and glancing at Jodi who was looking at him with her mouth open. Jodi's stare splashed cold water on his behavior.

As suddenly as it exploded Guy's temper dropped into a feeling of chagrin.

"My God," he said. "I've let my feud with glutinous bureaucrats destroy my common sense." He turned to Governor Aker. "Forgive me. I treasure your high offer, but my place is in Nevada. I've got companies which need undivided attention. President Roland should know better. He's got me up to my ears in problems defusing atomic weapons in the former Soviet Union. We're fighting Iran's high jinks together..."

Governor Aker listened to Guy unwind. Then he said with a wry smile.

"I understand. I don't think you've got the temperament for a U.S. Senator anyhow. But then, neither did Henry Ford, Thomas Edison, J.P. Morgan or John D. Rockefeller."

"Again, I'm sorry," said Ericson. "Only our congress has become so rotten I think it's like cancer, the only cure is removal by radical surgery."

"To be replaced with what?"

"A new body. Shaped by another constitutional convention. Junk congress just as our forefathers demolished the old Continental Congress."

Next morning at breakfast Guy Ericson was in good spirits. He could hardly keep his hands off his wife, who didn't seem to mind in the least. Governor Aker recognized the almost hypnotic influence Jodi had on her husband. Guy was humming his football team's fight song.

Tony Pellini winked at Governor Aker.

"What a difference a day makes," Tony sang softly.

"I had a talk with the President early this morning," said the Governor. "He gave me an idea I should have thought of myself."

"Yeah," said a disinterested Guy Ericson with a shrug. "What kind of idea?"

"On who to appoint U.S. Senator in your place."

"Suit yourself. That's your responsibility. Goes with the job."

"You're not interested?"

"I trust your judgment. Long as it's not me I'll back your hand."

"Promise. Without any tantrums?"

"I said I was sorry," Ericson barked in a bit of pique. "I'm not good at getting on my knees to apologize."

"He stood on his feet to propose," laughed Jodi.

"Well," said Governor Aker carefully, "with your blessing I've picked the next U.S. Senator from Nevada."

"Fine," said Ericson. "You've got my blessing. Who'd you pick? Some judge or legislator? I hope it's not another damn lawyer?"

"Sadly, it is." Governor Aker was enjoying his little game of suspense. Tony Pellini grinned. Guy was more bored than concerned. He studied the plush full mouth of Jodi and had an irresistible urge to drag her back to bed.

"The President suggested the best possible selection for the State of Nevada would be Jodi Ericson...I agreed."

Governor Aker and Tony Pellini drew back their chairs and waited for the bomb to burst. Instead they were confronted with total surprise and bewilderment. Guy and Jodi sat immobile, staring at each other in shock.

"In my campaign Jodi proved herself thoroughly familiar with Nevada and its problems with the federal government," Governor Aker hastened to explain. "There's just no other person who comes close."

Guy Ericson caught himself on the edge of another explosion.

"Such a decision belongs to my wife," he said, biting his tongue and fighting for control. "We agreed before our marriage to allow each to pursue independent paths. The very reason she continued to work for you, Governor."

"I'd have to think it over," said Jodi, easing the tension. "I certainly don't want to be away from Guy." At the same time she thought of her deep commitment to her father and grandfather. Love or revenge. It was a hard choice.

The next few days were agonizing. Guy was surprisingly sympathetic. He knew of her commitment.

"God puts us on earth to travel our own destinies," he said at last, putting the ball in Jodi's court. Guy knew if he opposed Jodi's opportunity she might later harbor resentment which could grow into a cancer, destroying both their love and marriage. He understood how desperately Jodi's pride wanted to return to Washington as a winning gladiator. Her humiliating exit still rankled.

"My jet can pick you up anytime," he graciously conceded on the third day of Jodi's agony. "Leave any noon and you can be with me for dinner and a full evening of togetherness."

On the fourth day Jodi told Governor Aker she would gladly take the job. That night she and Guy made love as if it were the last night on earth. Neither slept five minutes.

19. SENATOR JODI ERICSON

Senator Jodi Ericson had been in office six weeks when she got a call from Governor Aker.

"Got a friend with a problem," he said. She could see his freckle-faced smile over the telephone.

"Yes, Governor. How can I help?"

"Sudden heavy rains flooded my friend's land. It's been slow recovering. Now bureaucrats want to call his farm wetlands and take away his livelihood."

"They can't do that."

"Hell they can't," erupted the Governor. "The barbarians in congress passed laws which have little respect for private property."

"Barbarians?"

"Civilized people honor and respect private property. Theft is a barbaric act. But congress has licensed EPA, the DEA and other agencies to take at their own discretion."

"So what's the problem?"

"Remember the man sent to prison for filling in a few wet spots in his yard?"

"I remember," said Jodi. "A disgrace to our judicial system."

"The government throws a home-owner in jail, enacts retroactive taxes and singles out companies for regulatory harassment to satisfy screaming activist groups...who think all property rights are subject to their control."

Jodi had never known the mild mannered Governor to be so furious. His voice broke several times into upper octaves.

"I know," said Jodi. "A man back here asked a pertinent question. 'Is the government an honest person?'"

"A farmer named Gary Graydon suffered too much rain in a short time...soaked his soil and hasn't drained fast enough to suit the EPA. Now they classify it wetlands."

"Mental derelicts aren't all in institutions," said Senator Ericson. "I'll look into the matter."

"While you're looking, try this Hitler concoction."

"Another problem?"

"A monumental one for states and cities," fumed Governor Aker.

"EPA hatchet job, no doubt?" Jodi tried introducing a bit of humor to calm down the Governor. It didn't work.

"EPA's no laughing matter. Of the entire tax increase we were forced to swallow last year, 40 percent came from the EPA."

"Forced to swallow...?" Senator Ericson let the question hang in the air.

"Unfunded mandates," exploded the Governor. "Congress hasn't the guts to face the people with more tax increases so it writes laws which stuff the costs down the throats of states and cities."

"I thought congress under Newt Gingrich addressed that issue?"

"Addressed is right," declared the annoyed Governor. "The final watered down version barely got past the front door of the senate."

"Meaning they still left you with numerous rules and regulations without any money to carry them out?"

"Our self-anointed guardians exercised their usual irresponsible arrogance," exclaimed the Governor. "Like buying what they please then charging it to someone else's credit card."

"I know," said a sympathetic Jodi. "Many beneficial things exceed the cost an individual or nation can afford."

"And many not so beneficial," said Aker. "Like kids in a candy or toy store."

"Too many legislators have become like small children," agreed Jodi. "They want everything and cry and whine if they don't get it."

"Roger," said Governor Aker. "They want us to dig the money out of people locally to execute their stupid regulations."

"Like selling your house to pay for your own hangman?"

"Exactly."

"Then why do you do it?"

"Huh?"

"I said why do you pay? If I'm told to buy an airplane and I've got to sell my house to buy the damn thing, I refuse."

Governor Aker hesitated for a moment, stumped for an answer.

"Because it's the federal government. That's why?"

"It's also a government of the states who are supposed to have kept certain powers to themselves. In my opinion the bureaucracy flagrantly encroaches on those powers."

"You would defy the federal government?"

"Why not? Might wake them up. Who's to enforce the blatant dictatorship of congress?"

"The President."

"President Roland believes as you do. Thinks the Imperial Congress exceeds King Louis XIV. He's in control of the Armed Forces. So who's to make you collect from your own people to carry out mandates written by a congress with no knowledge of what they're doing?"

"The judicial system always backs any raise in taxes."

"Because not one judge in a thousand ever met a payroll, ran a business or did anything but draw his salary from the public tit...whether city, county, state or federal," said Jodi. "Time states threw down the gauntlet. To paraphrase Stalin, how many divisions congress or the judicial system got?"

"Whew," said Governor Aker, wiping his brow. "You throw some hard curves."

"Fast-balls. Hard fast-balls straight at the extortionists. Time for them to find out they can't get away with uncontested seizures any longer."

"Fine," said Aker. "That's well and good if backed by a President with guts."

"You know Roland. You couldn't have a President with more guts. Give him a ring."

"Far cry from recent White House occupants," snickered Aker.

The following Monday Jodi's husband called.

"You'd better make it home this weekend. These chorus girls are beginning to look pretty damn sexy."

"Oh, yes," said Jodi. "Keep your appetite under control. I'll be there. Then you can gross out like a male lion."

"I've heard they can do it 100 times."

"That's what an article in National Geographic claimed," Jodi said with a half smile. "Think you're up to it?"

"Maybe not 100, but right now I'd come damn close."

"You're no more eager than I am. I need a strong shot of that Viking adrenaline."

"Oh by the way," said her husband, "I've got a regular customer with a problem caused by that stupid government of yours."

Of mine, Jodi thought? Since I became U.S. senator, Guy and Governor Aker act as if she were the government. They regularly had lunch together, now they began to sound alike.

"What's the problem?" Jodi asked sweetly.

"A friend owns a single airplane and performs charter flights, fishing trips, sightseeing jaunts over Lake Mead, vacation excursions into Mexico and hunting trips into Canada and Alaska."

"Yes?"

"That stupid DEA nailed him for hauling a guy to Mexico with cash for a drug deal. They took his plane but let the guy with the cash off Scot free."

"Almost identical to action the DEA took against a man flying money into Canada. The pilot knew nothing. Was an ordinary charter flight. DEA confiscated the airplane from the innocent owner and let the drug merchant off free. Even letting him keep all his money too."

"Chrissake," roared Guy, making Jodi move the receiver away from her ear. "How can any government be so stupid?"

"Bureaucrats always crowd the limits of stupid laws handed them by congress," said Jodi.

"What can my friend do?"

"Practically nothing. But I'll talk turkey to the head of the DEA. Try to pound a wee bit of common sense into his operation."

"If that fails?"

"I'm on appropriations. Money talks. To government agencies a threat can prove stronger than a bribe. I'll see what can be done. Tell your friend to send me a letter of particulars."

"Great. But don't forget to come home?"

"I won't. Even if the DEA tries to confiscate your airplane for carrying explosive material."

Jodi no sooner hung up than she got a call from a national television show. Would the freshman senator like to appear on Marvin Pringle's weekly forum two Sunday's hence? Before thinking, she said, "I'd be delighted." The next day she got a call from President Roland. Events were racing ahead with dizzying speed.

When Guy Ericson heard of her TV engagement he sent Tony Pellini for detailed coaching. Guy didn't trust the media farther than he could throw the State of Nevada. He feared experienced talking heads would tear apart a novice like Jodi. He became visibly upset when he learned "Jam" Johnson, Speaker Joe Metzer and Senator Mordecai Meyer were also guests on the same program.

Late that same evening Senator Ericson got a call from an old acquaintance--Ali "Knuckles" Rahman.

"Hear you're going on national TV?"

Jodi knew she should hang up immediately. Perhaps his wounds had taught the brute a lesson. Her curiosity overcame her better judgment.

"That's right, Ali. What can I do for you?"

"You damned near killed me."

"You tampered with my reputation."

"Doctors say I could have died from a clot."

"It's possible."

"Now I've got a medical problem for you?"

"Yes?"

"I've hired couple Mafia hit men. Got plenty in New Jersey."

"To shoot me?" Jodi laughed. "You've always been a big blow, Ali."

"No. I wouldn't go to jail for you. They'll do to you what you did to me. Only worse. They're going for the kneecaps. Those beautiful gams of yours won't look so hot deformed."

"You're a sick bastard."

For some reason Jodi couldn't hang up. Like waiting for a judge to pronounce sentence after a guilty verdict.

"Men are professionals," said Ali. "They won't miss like the stupid amateurs on Nancy Kerrigan."

Ali's salvo of venom was the last straw. She'd heard enough.

"Drop dead, Ali. Sorry I didn't finish the job."

"Watch for the knee-cappers, bitch. They're coming," taunted Ali as Jodi slammed down the receiver.

Despite her bravo, an uneasiness crept up Jodi's spine. She understood Ali meant business. After all, her fit of vengance had maimed him for two months.

20. BREAKFAST WITH PRESIDENT

The week prior to her TV appearance reminded Jodi of University cram sessions before final exams. Guy distrusted the station to give Jodi a fair shake, particularly when he learned "Jam" Johnson, Speaker Albert Metzer and Senator Mordecai Meyer were to be guests on the same program. Marvin Pringle, moderator of the national network show, had a reputation for fairness but that didn't always protect his guests.

Guy Ericson's power and influence with Governor Aker and President Roland were beginning to pay off for Jodi. The President called her--Jodi suspected at Guy's suggestion--and invited her to breakfast. The invitation was not purely social, however. Roland was a no-nonsense President with a tough agenda for curbing the powers of congress. For better than a year he had been working on a comprehensive constitutional amendment to put a restraining bit on the unruly animal.

"Otherwise congress will continue to trample us to death with uncontrolled taxes," he said at one news conference.

President Roland found clauses in the Declaration of Independence which reinforced his anger.

"Congress, like King George, '*has erected a Multitude of new Offices, and sent hither Swarms of Officers to harass our People, and eat out their Substance.*' Also, he added that like King George congress '*has imposed taxes without our consent*' by refusing to submit increases to a vote of the people." He quoted those clauses from the Declaration of Independence frequently.

Senator Jodi Ericson was in her office with Tony indulging in a mock TV drill when her receptionist announced Speaker Metzer was on the line.

"Good morning, Mr. Speaker. This is certainly a surprise."

"I understand we are to be on Marvin Pringle's TV hookup this coming Sunday," the Speaker said without preliminaries.

"So I've been told."

"I called to make sure we make no blunders before the public eye."

"What do you mean, blunders."

"Anything which gives congress or individual members a black eye. The public scorns congress enough, we don't want to air any personal dirty laundry."

"You think I'd air dirty laundry?"

"I suggest we work together...bury the hatchet."

"I never carried any hatchet, Mr. Speaker. It was you and your assistant who buried one in my head."

"The past is past. I suggest we forget."

"Forget how you hounded me out of Washington?"

"You got a year's pay."

"How you prevented me from making a living?"

"A mistake."

"Forget your AA called and threatened to break my knee caps? Make me a cripple for life?"

"He had good reason. You damn near crippled him."

"So you know his actions in the restaurant?"

"I know what he told me."

"What he called me before friends?"

"Another mistake. You shattered his pride."

"Do you have Jewish friends, Mr. Speaker?"

A surprised and perplexed Metzer said cautiously.

"Of course. Many are stalwart supporters. Why?"

"Suggest they forget the holocaust. Get their answer. Then call me back."

Attempting to follow through on Governor Aker's problem, Jodi placed four frustrating calls to EPA. Various departments disclaimed responsibility, each saying it wasn't their baby. She ended finally with legal.

"Congress passed the laws, we're only executing them," said a snippy young voice. He then hung up.

Two days later Senator Jodi Ericson sat across the table from Howard Bloomberg, Administrator of the Environmental Protection Agency. It had taken a call from President Roland's office to set up the appointment.

Howard Bloomberg was a graduate of Yale law school and a Rhodes scholar--information he managed to convey within a few minutes of meeting any stranger. He wore a heavy black beard and was short and squat with hair sticking out over both ears. Jodi was sure she had seen him someplace before. Then she remembered:--a hairy short horned musk-oxen on the cover of National Geographic. He had a pleasant speaking voice.

"Thank you for coming," said Senator Ericson. "I couldn't reach you myself. You were always in conference or away from your desk. Since you never answered my calls, I guess your secretary invariably forgot to relay my messages?"

"I've been terribly busy," said Bloomberg brusquely, ordering a scotch and soda from the waiter. Jodi had an ice tea in front of her. Bloomberg was forty minutes late.

"Me too," said Jodi sympathetically. "As a freshman senator I'm on the Appropriations Committee. They assigned me, among other things, to the EPA subcommittee. Normally I wouldn't have much power but it so happens I'm the swing vote. Got to decide whether to cut $13 billion from the EPA budget." Jodi innocently sipped her ice tea, glancing around the Pub Restaurant. It was a hangout for congressmen and their staffs.

Howard Bloomberg almost spilled his drink.

"You mean you're involved with EPA's new budget?"

"Up to my ears," smiled Jodi. "I'm chairman of the subcommittee."

"My staff never informed me."

"How could they? I could never reach you?"

"I'll make sure it doesn't happen again."

Senator Ericson changed the subject, deciding to let the cocky young administrator think over his dilemma.

"That heavy black beard," she said. "Doesn't it itch?"

"Not after the first week." He was disconcerted by Senator Ericson's beauty. Her cool manners upset him. "You don't like beards?"

"Don't like or dislike them. Think in today's society with its emphasis on health and cleanliness, beards belong more on bushmen than intellectuals." She smiled disarmingly.

"I assure you they are clean," said Bloomberg, uncertain whether to take offense. Was she digging him or making naive remarks?

"Some say they capture dirt."

"Busybodies who don't know what they're talking about." Bloomberg finished his drink in one gulp and held up his glass for a refill. A waiter promptly complied.

"Dad's friends in the war grew beards overseas," said Jodi, twirling the top of her glass with her forefinger. "They had a devil of a time cleaning up when they got home."

"Meaning?" The blood was rising around Bloomberg's neck.

"Underneath the beards were oodles of hidden blackheads and pimples which took months to clean up. Had to use special soaps and ointments." Jodi shook her head as if truly concerned. "Faces looked like raw meat."

Thereafter the conversation stayed on neutral grounds; football, events overseas and the condition of various ski resorts. After lunch Jodi ordered decaf coffee. Bloomberg a snifter of brandy.

"Getting a Rhodes scholarship must have been quite a feat, Mr. Bloomberg," said Senator Ericson when the table had been cleared of all but the drinks.

"Indeed it was." Howard Bloomberg blossomed to attention.

"Understand that's quite a debating society?"

"Oh it is," said Bloomberg enthusiastically. "One of its primary goals is teaching people to think and speak on their feet."

"I've known six or seven persons from University who were Rhodes scholars. They told me the same thing."

"That they constantly engaged in great debates?"

"Yes."

"It's true," said Bloomberg. "We often changed sides, playing the devil's advocate and honing our debating skills."

"Switching sides frequently?"

"Oh yes. Professors drew propositions on the blackboard and we never knew ahead of time which side of the room would be assigned what?"

Jodi tapped her fingers on the tablecloth.

"Doesn't say much for learning fundamentals; like chemistry, math, physics, economics or..." Her voice trailed off into silence.

"You're supposed to know those things before you get there," Bloomberg said irritably.

"The Rhodes scholars I knew majored in political science, literature, history, journalism or law. Hardly your basic sciences," said Jodi.

"We majored in debate," said Bloomberg. "A damn fine discipline to acquire. Wouldn't hurt most congressmen."

"They acquire their fluency after years of hard practice here on the hill." Jodi signaled for the check. "Lunch is on me," she said. "I sought this meeting."

As an afterthought, as if suddenly remembering, she took an envelope from her purse.

"Oh yes. Governor Aker of Nevada has a problem with the unjust treatment of a citizen. The matter's explained in this letter from him. Please look into it and call me."

Howard Bloomberg was puzzled.

"This what the luncheon was about?"

Jodi smiled. "Not necessarily. I also wanted to meet the elusive Administrator of the Environmental Protection Agency. I still have a $13 billion decision to make before the end of the week. I've always thought small solutions often solve bigger problems."

A red-faced Administrator stood looking down at the serene senator as she signed the credit card bill. She lifted a twinkling pair of eyes to his frowning countenance. Her voice was soft but firm.

"Generals lose battles charging well fortified positions."

The following day Senator Ericson was able to phone Governor Aker that his wetland friend need have no further worries.

* * *

Breakfast with President Roland was in the Family Dining Room at the northwest corner of the State Floor in the White House. A silver mirrored centerpiece displayed on the dining table was fashioned by John W. Forbes in 1820. The President seated Jodi, taking over the job from the solicitous attendant. He placed her on his right, opposite the Louis XVI fireplace mantel on which sat a gilded clock. Made by Dubuc of Paris it was typical of French clocks made for American trade during the first quarter of the 19th century. Decorated above with an American eagle, the figure of George Washington stood alongside.

"Sorry I didn't meet you at the door," said the President, taking his seat. "Just had another session with Henry."

President Roland was referring to his Vice President, Henry Whitaker.

"Henry's sick, you know?"

"I heard he had cancer," said Jodi. "But I understood it was under control."

"Well, it's not. Now he's sick physically and mentally," he said, looking up irritably as the waiter served each of them a glass of orange juice.

Jodi remained quiet. She wanted no place in a quarrel between the President and Vice President.

The President softened his tone. "The strain's getting to Henry mentally. Lost his old fire. He thinks he's going to die so he wants to make peace with the world."

"Nothing wrong with that," ventured Jodi.

"Unless it's Chamberlain with Hitler, the Shah with Khomeini or the President with Congress. Gives the scoundrels breathing room...an opportunity to regroup."

"Sad to hear he's so ill," said Jodi.

"Me too. He's getting to be a frightened old maid. The Speaker and Mordecai are gleefully jumping on his condition to bully him into bad decisions."

The President reached over and took Jodi by the hand.

"Enough of my troubles. Guy is worried about his lovely wife on the upcoming TV show. Thinks she'll get shafted."

Jodi laughed. "He's found it easier to worry about me than face all those boards of directors."

The President grinned. "I told him not to worry. That it's the opposition who should worry. Governor Aker thinks likewise. He's seen you operate before cameras during his campaign."

"I expect it to be rough. Live TV crucifies mistakes. Neither the Speaker nor Mordecai tenders me any love."

"Wonder why?" The President mused aloud, sitting back and appraising the beauty before him. Abruptly he got down to the business on his mind.

"Let's talk about your TV show?"

"Fire away."

"Presidents are unique individuals," said the President, watching Jodi carefully. "Like a Dr. Jekel and Mr. Hyde, their true character only emerges after they become President. More often than not they don't even recognize the change themselves."

A puzzled Jodi kept silent.

"I'm trying to warn you. Presidents have ulterior motives for every action they take."

"Don't most people?"

"No. Not like a President."

"Aside from added responsibility," said Jodi, "what's so different?"

President Roland appeared to change the subject.

"I've talked to your husband and Governor Aker. Each thinks you've got a cantankerous nature."

After a moment's thought, Jodi smiled.

"I suppose I have."

"Both said that you love a fight." The President wiped the corners of his mouth with a napkin, avoiding Jodi's questioning eyes. He knew she was mystified at the direction of his conversation.

"Your husband claimed you didn't utter a kind word to him until he proposed."

"I think he got his facts twisted. He treated me like I was a thorn under his saddle from the first day we met."

President Roland lifted an eyebrow in amusement. He enjoyed this spunky charmer. He watched Jodi as she ate her cantaloupe. Unhurriedly. She ate to refuel the system rather than to eagerly devour the food like most guests. Her eating habits would not please fancy gourmet chefs or TV artists who went to great attempts to please the palate.

In repose Jodi's features were regular and relaxed. Yet the way she sat in her chair gave President Roland the impression of a coiled athlete sitting quietly on the starting blocks. At the sound of a gun she would explode into action.

"Dominant males normally attack any challenge to their supremacy," he said slyly. "They've usually got an instinctive smell for dangerous rivals."

Jodi snickered.

"It's their smell for females which gets them into trouble."

"Males don't fight females in the rut."

"Except sexually," smiled Jodi. "That's what they're all about."

"Civilization brought the feminist movement into more sophisticated ruts," laughed the President. "But you and Guy seem to have solved potential conflict by an excellent marriage."

"Mr. President, you claimed an ulterior motive for inviting me to breakfast?" She wished to evade any discussion of her marriage. It was superb by any count--physical or mental.

"Yes. I want to be brutally honest with you because my program will take guts--and plenty over the long haul."

Jodi stared at the President long and hard. What trap was he planning to spring?

"Mr. President, what are you trying to tell me?"

"I want to get rid of congress."

Jodi gasped. She felt a tightening in her chest and a shortening of breath.

"At least this congress," President Roland added. "Just as our forefathers got rid of the incompetent Continental Congress. We need a new type legislative body with controls."

"What type of body?" Jodi asked uneasily.

"One requiring an amendment to the constitution."

Jodi breathed easier. At least President Roland wasn't planning a coup of some kind.

"No question congress needs reforming," she said with an inner sigh of relief.

The President reached down at his side and pulled up several sheets of paper.

"In consultation with experts over the past several years I've drawn up some needed changes. Please read them and see if you concur. Some are programs Newt Gingrich and his friends tried to accomplish a few years back and failed. The old guard, like old communist guards in Russia and everywhere, gutted and diluted efforts until the nation ended back near the starting gate."

"I know," said Jodi. "Their brave new contract got sand-bagged with enough amendments to overload any horse."

"Very reason I insist on an all or nothing constitutional amendment," smiled President Roland.

"Like the parent who uses a belt for the kid's own good?"

"From your trail of speeches in Nebraska, Nevada and here, I've found little in my proposal you can't support."

Jodi took the sheaf of papers and glanced questioning at the President.

"Read them thoroughly in the next few days," he said. "Get back to me before the program. If an opportunity comes up to broadcast these ideas there's no better source of a trial balloon than your TV show."

"So I'm to be a guinea pig?"

"You might call it that," he smiled. "Or you might precipitate a movement to rescue the American people from a taxing octopus."

"I don't know," said Jodi hesitantly. "Newt Gingrich tried his damnedest. Now you give the job to a freshman senator on her first national TV hookup?"

"Don't force the message. Yet it's time we started dealing with issues on TV programs instead of name calling. I want to give you a backup of substance."

"Facing 'Jam' Johnson and Speaker Metzer I'll need all the help I can get."

"David had only a sling shot," smiled President Roland.

"David was an expert," said Jodi. "He knew his weapon. I'm a lamb facing professional knife throwers."

21. TV ORDEAL

Jodi entered the studio on the arm of a concerned Tony Pellini. Guy had violently opposed her appearing opposite such experienced political pros.

"Suppose the Speaker tries to destroy your credibility by telling a national audience he once fired you?"

"He knows better," said Jodi. "A rape scandal involving his office in today's climate could drop him into a political cell in hell."

Marvin Pringle, the host, was a throwback to David Brinkley. A solid newsman who gave his guests a fair shake. "Jam" Johnson, like Sam Donaldson, loved his own voice and interrupted often and loud. But Jodi felt he was a shallow thinker and more noise than substance. She did not intend to be intimidated by a buffoon.

"Jam" got his nickname during Viet Nam when he twice returned to base with jammed guns, or so he claimed. When nothing wrong was found with the weapons he was grounded and returned to the states. No court martial charges were filed but the nickname "Jam" stuck.

Speaker Metzer and Senator Meyer were buddies. Meyer was a subtle believer in the old McCarthy technique of destroying his opponent by innuendo--tossing a shocking rumor in the air and letting it fall into the tabloid swamps. No proof was ever offered.

When the cameras began to roll, Marvin Pringle introduced the guests.

"Jam" Johnson asked the first question.

"Senator Ericson, I understand you married one of the nation's wealthiest men and he bought your senate seat from Governor Aker."

The audience gasped at the effrontery and downright malignity of the remark. Jodi's face flushed red during the embarrassed silence which followed. She adjusted the microphone on her blouse but said nothing.

"Jam" Johnson leaned smugly back in his chair.

"Well, Senator Ericson, will you please answer the question?"

"I heard no question, Mr. Johnson. I heard you make a monstrous lie which would cause Goebbels to puke in your face."

Jodi had not intended to resort to such crudity. She regretted the remark instantly. She realized she had fallen into "Jam" Johnson's trap of trying to reduce her to an ignorant puppet. And he had just about succeeded with his opening foray.

"Jam" Johnson grinned. "My question is, do you think you have the background to be a U.S. Senator?"

"I think the constitution adequately addresses the requirements for a U.S. Senator," said Jodi. "I hardly think any extra input is needed from a man who fled enemy fire."

Marvin Pringle interrupted before an unwanted brawl erupted on his program.

"Another guest, Senator Mordecai Meyer, would like a few words. Please go ahead Senator."

"In Nevada, Senator, you claimed this body, which you later joined, was a 'temple of money changers'," said Senator Meyer. "Just what did you mean? And have you had second thoughts since coming on board?"

Jodi grasped the underlying reasons for his question. Was she going to be a maverick or go along with Senate leadership?

"My understanding always has been that our forefathers created congress to serve the people."

Senator Meyer nodded. "A task we address daily."

"No. A task congress rarely addresses," said Jodi. "Insensibly congress has become a body which exploits the people in every conceivable manner--supposedly for their own good. Like ancient surgeons bleeding the patient to make him well."

"A hideous comparison."

"But accurate," said Jodi. "Bleeding citizens through unchecked taxation eventually brings economic death to all."

"We tax the rich to..."

"Let's not start that class warfare again," interrupted Jodi. "Your cry for economic purity is as honest as Hitler killing Jews to purify Germany."

The Senator's face turned livid with rage.

"Our programs kill nobody."

"Except those small businessmen trying to realize the American dream. You strip them of their earnings then tell them 'now go out and create jobs.'"

"Our people want equality."

"Equality of opportunity. Not equal prison sentences."

"Tax money goes for good causes," said an angry Senator Meyer. "Our people demand universal coverage."

"Universal is camouflage for tyranny," snapped Jodi. "Hitler sought universal purity; as did the Spanish Inquisitors, Khomeini, Cromwell and a host of other bigots."

The Senator's voice rose.

"Congress was elected to serve the people."

"The President also was elected to serve the people," responded Jodi.

"It's up to the President, one person, to comprise with 535 members of congress if we're to get anything done," insisted Senator Meyer.

“Each was elected to serve the people,” said Jodi. “They were not elected to compromise with each other. The constitution went to great lengths to keep them separate.”

“A stupid game plan by the President can accomplish nothing.”

“If Congress and the President each hit on the same objective, then congress can forward bills which the executive can sign into law,” said Jodi. “If their paths do not converge in serving the people, more likely than not it’s a bad bill and shouldn’t be enacted into law anyhow. Gridlock is not evil, it merely says an idea is not sufficiently worthy to be imposed on all the people. We suffer from too many laws, not too few.”

“Without compromise we’d never be able to raise money for new programs,” declared Senator Meyer in a burst of pique.

Jodi smiled. "Like President Reagan said 'there you go again'. Bleeding patients to distribute blood to your own pet projects...so often strewn with failure."

Marvin Pringle stepped into the fray during a moment of silence.

"Our third guest is Speaker Metzer who joins us from his office in the Capitol." Pringle turned to the master remote.

"What is your response to the charge Congress is bleeding the people?"

"Depends upon your point of view," said the Speaker. He appeared jovial and relaxed in keeping with his image. Running his fingers through his unruly white hair like a comb, he spoke in the tone of a teacher addressing a wayward pupil.

"We think of it as administering transfusions. Taking from a blood bank to provide for those in need."

Jodi studied the Speaker's eyes. She knew his mobile dissembling features and pleasant smile created a permanent camouflage to his real thoughts. But his eyes, set back into his heavy face, were windows to his mind. At the moment they were hard, cold, and unpleasant.

Jodi said. "Speaker Metzer, donating blood has long been a voluntary American option. And it has been quite successful. But you extract bloody taxes without regard to the health of the individual...and without his consent."

"Pure hokum. Taxes are extracted to pay for social programs constituents demand."

Marvin Pringle addressed his question to Senator Jodi Ericson.

"What do you think is the single biggest problem with Congress?"

A small glint sparked Jodi's eyes then immediately vanished.

"Congress' inability to cap its greed," she answered without hesitation. "There's nothing in the constitution which requires it to do so. It can tax, tax, tax and tax until revolution strikes and our unique form of government is forever lost."

"Nonsense," said the Speaker. "Doomsday talk from fanatics. Congress rejected a balanced budget amendment under my predecessor, Newt Gingrich."

"Engineered by the senate old guard," said Jodi.

"A lid on taxes wasn't practical. It didn't allow for emergencies."

"A perpetual condition facing congress," said Jodi. "Under the Czars landowners extracted one-third from their serfs. In the U.S. today government takes fifty percent. This trend makes a Czar look like a savior."

"Voters can always vote us out."

"With gerrymandered seats, franking privileges, PACs, and perks coming out your ears? Pure nonsense."

"Colleagues are pushing term limits," said the Speaker. "Many states already have adopted them."

"Short-sighted idea. With no limit on its power congress can still raise taxes, grant itself 100 percent pensions then move in and take top jobs in the bureaucracy." Jodi laughed, a long deep-throated chuckle. "A better idea would be to require taxes to be submitted to a vote of the people. And term limits abandons power to the bureaucrats. Like leaving prisons in charge of the inmates."

"Jam" Johnson, inwardly raging at Jodi's reference to his Viet Nam record, had been quiet long past his normal time.

"The media has kept democracy alive."

"The media has created bigger problems than it has solved," said Jodi, barely able to hide the disdain she felt for “Jam” Johnson.

In Jodi’s campaign work in Nebraska and Nevada she had experienced the leaks, anonymous sources, blatant prejudices and outright falsehoods exploited by the media. Character damage was of no concern to many of them.

Marvin Pringle questioned.

"Just how would you handle fiscal responsibility, Senator Ericson?"

"Jam" Johnson and the Speaker nodded approval of the question.

"A constitutional amendment is the only answer," said Jodi, warming to her subject. "Total federal spending not to exceed twenty percent of the Gross National Product..."

"About taxes?" Pringle asked.

"Congress should not raise taxes unless by two-thirds vote of both House and Senate and signed by the President," she said, adding, "except following a declaration of war by the congress or declaration of National Emergency by the President."

"That would practically immobilize congress on needed new taxes," the Speaker protested.

Jodi smiled. "The general idea."

Marvin Pringle was delighted at the way the TV show had erupted. He could tell from the studio audience and the attention of the workers that the show was having an impact. People were interested in substance.

The underlying animosity between the participants was better than a murder trial. Ratings would go up.

"Any more suggestions, Senator Ericson?"

Jodi smiled, waving a manicured forefinger at the moderator.

"Trying to get me hanged from the nearest yardarm?"

"No. Only to surface new ideas. Our listeners might like them."

"How about reducing spending twenty percent each term until we get a balanced budget?"

"An impossible dream," laughed Pringle. "How would you do that?"

"Cut congressional salaries twenty percent each term until a balanced budget in real dollars is achieved," smiled Jodi. "I'll bet members would pull pork barrel tapeworms out of the bureaucracy so fast fish in the Potomac would begin to look like whales."

"Anything else?" Marvin Pringle glanced at the glum faces of his neglected guests as the audience roared approval of Jodi. This beautiful woman who sat nervously crossing and uncrossing her legs was a camera man's delight.

"Yes. There's more," Jodi said. "Sure you want to hear?"

"That's why we have this program," said Pringle. "Let listeners form their own opinions."

"No federal agency, except those mandated by the constitution, should live beyond 18 years without being recreated by law," said Jodi, giving Pringle a skeptical glance. "An independent task force should conduct a thorough re-analysis of an agency's worth every 18 years. Keeps bureaucratic operations from drawing money and hiding in budget cracks year after year."

Speaker Metzer determined to be heard.

"We can't progress if we're going to nit-pick everything that's been done in the past," he said. "We need to move the country forward and create change."

Jodi's face twisted into a skeptical smile.

"My dear Mr. Speaker," she said, "I think I'll get sick on national television if I hear one more time from leaders who are magically going to move the country forward or create change with the wave of an arm."

She studied the shocked Speaker with amusement before continuing.

"Congress has created enough change. You create change in an individual's economic status when you foreclose on his house or farm or when you repossess his automobile. That's real change, the kind you've given us in the past.

"You can move forward into an enemy ambush and get everyone killed. Ask any soldier. Or move forward into quicksand or in front of a runaway truck. Neither particularly wise decisions for change."

"You're twisting my thoughts," fumed the Speaker.

"No. Advising you to drop that same idiotic worn-out appeal to voters you think aren't quite bright," said Jodi. "If you really want to help citizens, be specific. Say exactly what you're going to change. Then live up to it. Saying you're going to improve their lot by spouting false promises of change on television, then sending them a bill from the IRS is not the answer. The public's had its fill of lies from congress."

Speaker Metzer was determined to nail Jodi as a zealot or revolutionary. Politically wise, he knew those outside the main stream inevitably were dismissed as kooks.

"Senator Ericson," he said disarmingly, "isn't what you're really saying: our form of government stinks and you would like to overthrow it?"

Jodi recognized the harpoon thrown at her heart. She studied the Speaker dispassionately for a full minute.

"You'd like me to bare my soul, wouldn't you, Mr. Speaker?"

"If you have one? Yes."

Glancing at the audience, Jodi took her time to gather her thoughts.

"We constantly read of clear and present dangers from foreign sources," she said. "We fear the action of unprincipled nations. Yet outside forces rarely destroy a nation."

"Hookum," snorted the Speaker. "Read your history."

"It's the rot within," said Jodi, ignoring the Speaker and addressing an audience captured by the electric animosity flowing between the two.

"The Roman Empire eventually destroyed itself by the rot within--Emperors who never bothered leaving their palaces. Rot within the Spanish Empire and it's Inquisition and demands for a universal ideology reduced it to a whimpering monarchy. Soviet internal rot eventually brought about its collapse as well as the fall of numerous satellite dictatorships..."

Speaker Metzer interrupted. "Napoleon and Hitler certainly weren't defeated from within..."

Jodi responded tartly.

"France and Germany's internal rot drove the people to accept a Napoleon and Hitler. Wellington didn't defeat Napoleon. He defeated a France so decayed from years of kings running a corrupt bureaucracy that it could no longer sustain itself. Germany was an economic disaster primed for a Hitler, as was Iran primed for Khomeini."

The Speaker tapped his finger to his forehead as if Jodi was off base.

"So what's that got to do with us?"

"So we're heading down the same road of internal rot brought on by a taxing and spending congress which has no more limit on its taxing and spending excesses than kings, emperors or dictators," said Jodi. "We're fast becoming a nation of Sodom and Gomorras. Too many believe the good life is their right, no longer a goal to be achieved by the same route as our ancestors--work, saving and responsibility. Get a megaphone and demand food, clothing and other goodies from politicians as a right. Then a congress, greedy for votes, can match the greed of the noise-makers.

"Keep this up," declared Jodi, throwing caution to the winds, "then it's back to another Hitler, Stalin or new messiah who will take away every last vestige of freedom. For as Justice Holmes said, given the choice between impoverished anarchy and order, people will always choose coercion."

Marvin Pringle, at a signal from the control booth, cut off the choked-up Speaker's retort.

Speaker Metzer ground his teeth in silent rage. Jodi Ericson was driving a deadly harpoon into the heart of his congressional power. She could no longer be tolerated as a simple odd-ball maverick. She must be destroyed before she smashed him and his meticulously constructed power base.

22. VEGAS

Jodi walked to the window and looked across at the mountains. She stood naked with her back to Guy.

"I'm exhausted. Riding rodeo and roping calves wasn't nearly as tough as riding a Viking."

"Seas too stormy for you?"

"Never. I love it."

Guy looked at her nude body and his mouth became dry. Jodi was designed for a sculptor's art. Underlying muscles rippled across her broad tawny back and down the sides of her rib cage. Long legs with smooth tanned skin glistened with sweat as she stood rocking heel to toe, a tigress waiting to spring.

Guy marveled at the circumstances which brought him this wondrous animal. He shuddered as he remembered how he had nearly missed the call. If it hadn't been for the assassination attempt on his friend President Roland, Guy knew their marriage might never have happened. Their independent ways would have thwarted any merger. Only the shock of seeing Jodi lying in a pool of blood brought him to the realization that some perverse smoldering love had accounted for his belligerent behavior. It had been a false form of defense, fearing to get involved with another woman after his disastrous first marriage.

Guy sighed as he walked up to Jodi. Perhaps the gods did know what they were doing?

"I love you," he said, his voice low and strained.

"I know," she said without turning.

He placed his arms around her and held her breasts firmly in each hand.

"I never thought it could happen," he said.

His touch was an electric current tingling the skin all over her body. Jodi never had experienced anything like it. At first she had fought the feeling, not wanting to be a slave to sex. Now she gloried in abandonment, unashamed and eager for the wildest maneuvers. Each new love session did the opposite of what she expected. She originally thought these trips back to Vegas would curb her appetite and satisfy, at least temporarily, sexual demands. Instead, like an addiction to drugs, each encounter demanded more.

"What could never happen, Guy? Loving me or loving another woman?"

"Neither." He moved against her, feeling the ecstasy of her warm skin against him. "When I met you I was like a gladiator who had barely escaped from the arena, mauled but alive. At the bottom of the barrel, emotionally drained."

"I thought, what an arrogant snob," she said, holding his hands tight against her breasts.

Guy laughed softly.

"A protective device. The greater the danger, the higher the protective fence."

"You thought me dangerous?"

"Not consciously. But instinctively. An internal warning," he said.

Jodi chuckled from deep in her throat. Guy listened to the soft musical sound. The clean fresh earthy aroma that wafted into his nostrils from Jodi's exercised body reawakened his desire.

"How right you were," she said. "Now I've become downright deadly. I expect you to pound me into ashes with your love, leaving not a single match-stick of fire for another woman."

"I'll have to get in better condition?"

"Don't worry about that. You're lean and hard enough. I'll just have to give you enough workouts to keep you that way."

"Good. We can start by you coming back to bed." He let go one breast and pulled her around to face him with the other.

Then it started all over again as if it were the first time. Savage and wild. But to Jodi each succeeding time was always better than the last.

Insatiable was the only word to describe it. Guy carried her into another world. One where pain combined with frenzy produced ecstasy. He invaded every niche and crevice in her body with his own powerful current. Beads of sweat popped out from her skin as his fever possessed her. She was rushing headlong into a black hole of oblivion.

"Guy! Hold onto me."

She bit her lower lip hard to keep from screaming as she imagined herself swirling helplessly towards the black hole. "Oh, my God. Don't let go," she cried aloud, just as she felt his hot surge of life blast its way deep into her body.

It was a full five minutes before she opened her eyes to find Guy staring down at her. He still hadn't removed himself from her body.

"I think you should come home more often," he said.

Jodi could only nod in helpless agreement.

23. LONDON CONFERENCE

The months flew by swiftly as Jodi settled into her job as U.S. Senator. Senator May Trudeau moved closer as a friend and supporter. They made quite a pair. Two sex goddesses from the same state. A media bonanza. There was one big difference. As a solid environmentalist Senator Trudeau was popular in both Houses. But Jodi, constantly waging President Roland's war for congressional reform on TV and in the news, incurred the fanatical enmity of Speaker Albert Metzer and Majority Leader Mordecai Meyer. She was attacking their personal domains and they had no choice but to destroy her before she destroyed them. President Roland recognized the heat she was taking in his cause.

"Civilizations which lasted had one thing in common--control," the President told Jodi. "The present congress is a babble of voices. Loud, begging, whining demands for more and more programs to appease the irresponsible. To govern successfully one can't always listen to the anarchy in their legislative assemblies."

President Roland offered one more bit of wisdom when he invited her to accompany him on Air Force One to visit her husband and Governor Aker.

"A President who doesn't realize Congress in 200 years has always cut and run when hit by real pressure, is a naive chief executive unqualified for the job."

"My father pressed me with one admonition," said Jodi.

"Yes?"

"Play it right the first time because the present quickly becomes the past--never recoverable."

*   *   *

With her husband Jodi gradually became immersed in the President's labors to defuse the dangerous residue of weaponry left from the breakup of the Soviet Union. The President continued to pour money into the effort through MRI, Inc.

Late one Friday afternoon Guy called Washington to see if Jodi would like to meet him in London. He planned to attend a conference with the President and top scientists the following Tuesday. Vice President Whitaker originally was to chair the event but he was back in the hospital with a flare-up from his cancer.

"Oh, yes," Jodi answered without hesitation. "I'd love it."

Getting away from Washington into a new environment was like a gift from a Genie.

"Great. I thought we'd make it a second honeymoon," said Guy.

"We never had a first," Jodi said. "But I'll not quibble. Your timing couldn't be better. I just bought a new negligee from Frederick's."

Guy laughed aloud into the telephone.

"Wasted money. You'll never get a chance to wear it. I haven't seen you for three weeks."

"Perhaps I can wear it afterwards?"

"For dinner? You'd create chaos in the dining room."

Jodi shivered. Thinking of Guy still did things to her.

"Where in London?"

"I've got a suite at the Churchill Hotel. I'll arrive Sunday afternoon. I'd pick you up but need to clear with United Nation bureaucrats in New York first. Prepare yourself."

"Oh, I will," Jodi said, slowly cradling the phone after hearing a click on the other end of the line.

*   *   *

Five minutes after entering the suite at the Churchill in London Jodi and Guy were in bed together.

"Most fun one can have without laughing," Jodi said, without apology for the old cliché. She lay on her back resting after a second vigorous round of love-making. "I thought male stamina was supposed to peter out first."

"Not Viking vigor," said Guy. "They performed with captive maids on open rolling seas in boats not much wider than this bed."

"That's dexterity. Not stamina."

"My ancestors had both," bragged Guy with a satisfied smile.

He lay on his side stroking Jodi's hair, reveling in the scent of her body and the touch of her smooth skin. He ran his fingers around each breast, stopping to test the firm nipples with his mouth, then proceeded over the belly, down the inside of her thighs to the arches of her feet, before returning to cover her mouth with kisses. Her full lips were like insatiable honeyed flowers which trapped victims into deep deadly pits. More was never enough.

Guy lifted his mouth from a breast, giving it one final flick of his tongue.

"For years I've been screwed by politicians, but never did I dream I'd get such pleasure in screwing a U.S. Senator."

"Nor I in seducing an arrogant billionaire," Jodi laughed. "Like the old saying, 'where were you when I needed you?'"

"You need my protection now," said Guy. "We're dining with the President. He's developed a fondness for you."

"Paternal only."

"Where you're concerned, I'd suspect the Pope."

"You don't trust me?" Jodi pushed up his nose with her forefinger before kissing him and biting him gently on the lower lip.

"Caesar trusted Brutus, didn't he?" Guy thrust his feet over the side of the bed. "You going to scrub my back, or do I have to call the maid?"

"Talk like that may get you scrubbed with a wire brush and lye soap," Jodi said, slipping past Guy's grabbing arms on the way to the shower.

"Where are we meeting the President?" Jodi asked. She sat before the vanity putting the final touch on her makeup.

"In his suite on the top floor."

"Not in the main dining room? What a shame. It's such a beautiful room."

"But a security pain for the Secret Service. Nor any fun while the world watches you eat."

"Anyone else joining us?"

"Two top scientists from Material Resources International. Nikita Alexander and Leo Fedor."

"Your company?"

"Yes. Funded by the U.S. to dismantle the weapon threats from the dismembered Soviet Union. President Roland wants to hear from the horse's mouth what's going on."

"The job certainly puts thousands of Russians to work."

"I hope they appreciate it," said Guy. "He got the program funded for one billion despite wailing and caterwauling by many in congress."

"Getting rid of atomic weapons costs less than war."

Jodi crossed the room towards the closet in a pair of tight fitting white panties and no bra. Her breasts jiggled to the rhythm of her walk. Guy flicked his lips with his tongue.

"You're going to have to dress in a separate room or be prepared to jump back in the sack every fifteen minutes," he said.

"I prefer the sack. I'm sure you'd holler *uncle* long before me," Jodi smiled, tweaking his nose as she passed in front. "How much time we got?"

"Not enough," sighed Guy, looking at his watch.

"Too bad." Jodi loved teasing Guy. His banter was like an infusion of fresh blood. Just having him in the room buoyed Jodi's spirits. She wondered what possessed her. She shamelessly realized she'd eagerly skip dining with the President to rousting in bed with Guy.

"Incidentally, your friends Metzer and Mordecai are here in the same hotel. Checked in just before you with the Speaker's gorilla, Rahman."

Jodi's spirits crashed from soaring ecstasy into a swamp.

"They're no friends of mine. They hate me."

"I know. They'd like to atomize you into thin air."

"They're creeps. They fight every thing the President tries to do." Jodi was incensed. "Wonder what they're doing over here?"

"On their way to Paris. Meeting unsavory friends from the Mideast. Those violently opposed to the President's policies."

"Isn't that illegal?"

"Called fact-finding trips. Also my sources checked the airlines. The Speaker is stopping for two days in Switzerland."

"What for?"

"Who knows. Probably to meet another anti-American? It's a neutral meeting zone for malcontents."

"How come you got so interested in the Speaker?"

"Because of rumors my agents uncovered accidentally about you. The Speaker and his AA hate you with a passion."

"You know the reason for that. Goes back years."

"But now you're in congress. Doing them a lot of damage with your assaults on their power," said Guy, pinching his fingers worriedly against his forehead. "I started tracking the Speaker mostly as a whim because of his past treatment of you. Now I think the man's a vicious animal beneath that fat charm. I've no intention of letting him harm you."

Jodi glanced at Guy and saw the care in his eyes. She had a sudden upwelling of love which nearly choked the breath out of her. To be the object of such love made her whole body tremble with wonder. How could it have happened? It frightened her. Surely such paradise couldn't last.

* * *

President Roland greeted Jodi and Guy affectionately.

He smiled at Jodi. "I'm about to nominate you to the Hall of Fame as my favorite quarterback."

She looked up quizzically as he led the pair to a conference room at the end of the hall.

"I'm not sure what you mean, Mr. President?"

"Pitching the ideas of reforming congress on TV, radio, in the news and in speeches across the country. You've been doing one helluva job."

"Ideas I believe as firmly as you. Otherwise I'd not be so zealous. I've certainly made enemies of our congressional leaders."

"A condition which proves you must be 100 percent right. If I thought the Speaker or Majority Leader supported you, I'd know we were on the wrong track," said the President.

Guy Ericson lent a disquieting note.

"I think Metzer and Meyer mean real harm to Jodi. Not just political."

Nikita Alexander and Leo Fedor were waiting at the entrance to the conference room. Jodi had not met them before. Nikita Alexander was a tall blonde American MIT graduate with Russian parents. He spoke Russian fluently. Leo Fedor was a graduate of the Air Force Academy who had gone directly into the space program. He was a Norwegian from North Dakota. Both now worked for Guy Ericson as top scientists in Material Resources International, Inc.

After a fifteen minute briefing on the status of the dismantling of the most dangerous weaponry, President Roland questioned Nikita.

"Any big space stuff escaping to Iran?"

"Not sure, Mr. President," said Alexander. "But big money talks. Remember the German secret service seizing 6 grams of plutonium 239 from an entrepreneur smuggler?"

Fedor added. "And the German sting operation which netted 300 grams of plutonium bootlegged from Russia?"

Alexander said. "It's rumored the smuggler was offered $100 million for his six grams--an amount no bigger than his thumb. He couldn't carry that much gold out with a strong mule."

"Countries willing to pay such sums for frightful weapons of annihilation shows we're a long ways from international peace," said Ericson. "Iran is also managing to buy all sorts of battlefield arms."

"Mostly for ground troops?" remarked the President.

"Yessir. Your actions are making it damn tough for them on what they really want--atomic and space capabilities."

"I'd really like to throw a tighter embargo around Iran," said the President, "but the fractured United Nations won't go along."

"Iran's mullahs now hate the U.S. more than they did under Khomeini," said Fedor.

"We're the ones tightening the screws," said Guy.

"Right," said Fedor. "Tehran mullahs particularly hate you, Mr. President, because you've been the most successful. Iran seeks to create a world Islamic empire and weapons of mass destruction lie just across the border. An easy sneak, except for your actions."

"A weak fence protecting crown jewels," said the President.

"They've learned it's far quicker and cheaper to buy arms than import raw materials, set up factories and train scientists. A ten-year job at best," said Alexander.

The President listened attentively to a discussion of Iran's third-world purchasing methods and its dealings with old hard-line Soviets who gobbled up cash for their own pockets.

"We need to move fast against Iran," said President Roland. "It's highly unlikely the Soviet republics will survive as democracies. When leaders start losing their grip, they'll be ripe for deals trading Soviet weapons for cash."

"You're absolutely right, Mr. President," said Fedor. "Without foreign assistance, Iran can get nowhere with their military modernization."

"Why is Iran such a rogue?" ventured Jodi.

President Roland smiled. He sat at the head of the rectangular mahogany table and glanced at the participants.

Jodi was like a fresh spring rose. An open white blouse topped a favorite purple skirt which set off her tanned legs. Her eyes crinkled with humor. She was an intellectual sponge. During the past months she had grabbed his ideas on reforming congress and added brilliant suggestions of her own. Like a bird dog seeking prey; once settled into a point, there was no indecision.

"Iran is a threat to world peace because it's driven by fanatic Islamic fundamentalists," said the President. "They believe the west is populated by vermin infidels who should be exterminated."

"Each year our work becomes more difficult," said Fedor. "A steady drumbeat of potential terrorism is gradually erupting. Ethnic and religious hatreds are opening old raw sores. Security barriers in hostile lands collapsed with the old Soviet border patrols."

"I know," said the President. "Dealing with deranged minds is futile. A lesson the UN seems incapable of learning."

"I remember Sheik Omar, that unclean looking spectacle finally picked up by the FBI on the bombing of the World Trade Center," said Guy Ericson. "Omar was quoted as saying 'Muslims must kill the enemies of Allah, in every way and everywhere, to liberate themselves from the grandchildren of the pigs and apes who are educated at the tables of the Zionists, the communists and the imperialists.'"

"Even Stalin didn't cast such a deep and wide net of hate," said Fedor. "We know from underground rumblings at our weapon sites that young Hezbollah terrorists soon will try a spectacular act to gain direct access to Allah."

"Until legislators deny entry to all immigrants whose passports are not first scrutinized by our embassies overseas, terror will grow in the U.S.," said Alexander.

"That's what we got embassies for," said Guy. "The surplus paper pushers now in the State Department could easily handle the additional job of security."

"But they don't, and won't," said the President. "They wait until the immigrants land, then we're left to corral an exploding field of rats."

24. PRESIDENT'S RUNNING MATE

President Roland leaned back in his chair in the oval office and touched the tips of his fingers together in meditation. An open convention would provoke suppressed wrath and create infighting which would tear apart party cohesion. He had to move. A decision couldn't be delayed much longer. The convention was only weeks away. Vice President Whitaker declared unequivocally he had no desire to run again. His visits to Walter Reed for cancer treatment were now on a weekly basis.

The intercom rang. His secretary sounded apologetic.

"Speaker Metzer and Senator Meyer are here to see you, sir. They have no appointment. They did call earlier to see if you were in, then hung up."

"Never mind. Send them in." President Roland got up and walked to the windows looking out over the Rose Garden. They undoubtedly wanted to raise hell again about his treatment of Iran. He understood the Speaker's position. He had numerous Mosques of well-organized Muslims in his district adjacent to New York. But Mordecai had no such problem. What was his motive?

President Roland greeted them each with a handshake and motioned them to a seat on the sofa. He sat in a wing-backed chair.

"What can I do for you today?" he asked pleasantly.

"We'd like to talk to you about the Vice Presidency," said the Speaker.

"Oh," said a surprised President.

"Since it's common knowledge Whitaker will withdraw sooner or later, we'd like to propose a candidate before opposing camps solidify into uncompromising positions."

"I've given that a great deal of thought," nodded the President. "Yet I've still been unable to come up with anyone to replace Whitaker. Perhaps partly due to a reluctance to admit Whitaker's condition."

"We understand," said Senator Mordecai Meyer.

"So our timing is fortuitous?" said the Speaker. "You've not selected a running mate?"

"Correct."

"We've got an excellent candidate for you. One who should strengthen your position at the polls," said an enthusiastic Speaker.

The President glanced at Speaker Metzer skeptically. If ever anyone brought a Trojan horse into the President's camp, it would be the Speaker.

"Who's this vote-getting wonder?"

"You know him well," said the Speaker. "Congressman Jesse Rowan, black chairman of the House Judiciary Committee."

The President stared at the Speaker as if he had fallen out of a tall tree. Congressman Rowan, leader of the black caucus, had been one of the President's most formidable foes. A tax and spend die-hard who had never seen a tax he didn't like. Despite being of the same party, Congressman Rowan was at the other end of the President's spectrum. Strict party loyalty under Speaker Metzer had become a thing of the past. Today there was greater diversity within the party than between Republicans and Democrats during President Clinton's term. Ross Perot had initiated the split. President Roland reflected, as he surveyed the members before him, that before long there had to be a complete realigning and renaming of parties.

"Before you categorically say no, Mr. President," Senator Meyer interposed, noting Roland's forbidding expression, "think of his many pluses."

"Yes, Mr. President," said the Speaker. "The time has come for recognizing change...time for a black. Jesse has 100 percent support of the black caucus; of black leaders and mayors throughout the country; and the blacks have solidified into a huge voting bloc."

"Black mayors throughout the U.S. were elected with white votes," thought the President. "Yet when Jesse Jackson ran for office 98 percent of the blacks gave him their votes. Few blacks crossed racial lines to vote for white candidates. Whitey was cut out completely. A one-way street."

The President studied the Speaker who could scarcely contain his ebullience. What was he getting out of this deal? Suddenly he realized Congressman Rowan had been loudly supportive of the Speaker's Iranian policies.

"Congressman Rowan," the President said, his fingers rapidly tapping the arm of his chair to relieve his annoyance, "advocates open immigration, bilingualism in education, more money for day care, more money for welfare and food stamps, more money for inner-city housing, and billions for training and retraining workers...For what? For jobs which don't exist? We need to reduce taxes so the small businessman can hire workers."

"People love change," said Senator Meyer. "I think we need a fresh face."

President Roland realized arguing with established convictions was a waste of time.

"We certainly need freshness, Senator," the President agreed, looking at his watch and rising to his feet. "Rowan's candidacy comes as a complete surprise but I'll give him serious thought."

"Not much time," said the Speaker.

"I know. You'll have my decision within ten days," said the President as he courteously escorted them to the door.

* * *

Regularly the President traveled back and forth to Walter Reed hospital. He wanted to sit down and exchange views with one of the few people in Washington who had no ax to grind or favor to seek--his Vice President, Henry Whitaker, now a permanent resident of Walter Reed.

Tuesday morning Jodi got a wakeup call from the President.

"How about a return breakfast?" he said cheerfully.

Jodi looked at her alarm clock on the side stand. It was 6:15.

"I guess you never sleep," she said.

"You've done such an excellent job of promoting my program in the news and on TV that I've got more work for you," he said jovially.

"I can hardly wait," Jodi said, dragging her feet over the edge of the bed.

"Is eight o'clock too early for you? It's a late breakfast for me."

Jodi sat bolt upright, finally realizing she wasn't dreaming.

"Not at all, Mr. President. Eight o'clock's fine."

"Excellent. See you at eight." The line went dead.

* * *

Breakfast was in the same room as before. The Family Dining Room. Painted yellow walls with white woodwork and a white ceiling. A white eagle stood sharply decorative against dark greenish marble. This time the President was waiting with a broad smile.

"I enjoy getting people out of bed," he said. "Guess it's because so many people seem to have no compunction about routing me out of the sack."

"No problem. I'm used to rising early. My dad had me out doing chores by this time."

"So he's responsible for hammering that figure of yours into shape?"

"Partially," said Jodi, her face flushing with embarrassment. It was the first time the President had ever made reference to her looks. "Sitting a horse all summer long doesn't allow the accumulation of many extra pounds."

The President continued the light banter until they had finished breakfast and were drinking coffee.

"You've been quite a hit on TV shows and with the public," he said. "My spies tell me you're pretty quick with the repartee, too. Especially when some bloke like "Jam" Johnson tries to bury you in a side-pocket."

"Like his predecessor, Sam Donaldson, he's mostly being wrong in a loud voice," said Jodi. "I wish all antagonists were such easy prey. He's like roping a tied calf."

The President laughed. "You do have a way of expressing things."

"Ranch lingo. A real problem when I got into University. Teachers almost despaired."

The President studied the raven-haired creature before him. She'd been increasingly on the cover of numerous women's magazines. There was a mysterious quality about her which even the slicks had trouble capturing.

"Well I've got a real job for you," he said. "One that'll make you think riding broncos was a merry-go-round."

Jodi looked up in surprise. She had been patiently wondering what chore the President had next. She had practically become the right arm of the President in the Senate and around the country in exposing the shortcomings of congress. In many ways it had been a lousy job. Foes like the Speaker and Mordecai Meyer tried to cast her as a simple-minded robot responding to buttons pushed by the President. Yet as she often discussed with her husband, only through a President could congress be curbed of its greedy grasping ways.

Jodi sighed, thinking to herself.

"Damned if you do; damned if you don't."

She said aloud, a bit fretfully.

"I've been trying to do a real job."

The President caught her discomfort and laid a solicitous hand on her arm.

"I know, you've done a superb job. That's why I want you as my Vice President in the coming campaign." He sat back and smiled as he watched Jodi try to recover from the shock.

Jodi gazed at the President, then at the mirror above the clock. Her eyes glazed as if blinded by a sudden burst of light. After a full minute of digesting his remark, she said.

"You're not serious, I hope?"

"Deadly serious. I've talked it over with both the Vice President and your husband. We're all positive you can handle the job as well as any Maggie Thatcher."

"And Guy agreed?" she said incredulously.

"Enthusiastically."

"The traitor," Jodi said, shaking her head in disbelief.

"Guy's not to blame for keeping it secret. I made him promise. There was the normal routine of checking you out with the FBI and CIA."

"Members of congress hate my guts. The Speaker and Majority Leader would like to see me walk a pirate plank."

"But not the public," said President Roland. "Polls rate you exceptionally high. The very reason Whitaker is hanging on until after an election. If I tried to appoint you, I'd never get a vote of approval from either the House or Senate."

"Hardly a consolation heading into an election?"

"Congress doesn't elect a President or Vice President. If they did I can name a dozen presidents who would never have survived the first ballot, starting with Abraham Lincoln and ending with Ronald Reagan and myself."

"What can I bring to your Presidency?"

Jodi knew one could be crucified in nasty debates. As a youngster she had watched the vicious attacks on Dan Quayle and Supreme Court nominees. It was not an arena to jump into lightly. She thought of Roman gladiators who had come out maimed even when victorious.

"You've had more campaign experience than most Vice Presidents," said Roland. "Congressman Pace said you got him over the hump in Nebraska. Governor Aker swears you won his election in Nevada. You've faced numerous TV and media antagonists, all loaded for bear. Most important, you've got my confidence," he added with a smile.

"A woman Vice President?" Jodi mulled the image over in her mind.

"The time is ripe for a woman on the ballot. There's clamor for a fresh face."

"A woman Vice President was tried once before," Jodi reminded the President. "Results were disastrous. Voters don't even remember her name."

"Unrelated problems doomed that effort."

"I'd not want to be the one to sandbag your bid for reelection. They loved Henry."

"Henry isn't up to the rigors of a campaign and wants out. He promises to support you in the major cities where he's strong."

Jodi looked the President squarely in the eyes.

"I'm not being coy. Just flabbergasted. If you really think I can help you, I'd be delighted to be your running mate."

"Not only help me, but help the nation," said President Roland. "I'm not picking you because you're glamorous, but because I think you've got the guts of a Nebraska meat-packing plant. A quality you'll need to face down a congress led by a modern day Boss Tweed with Robespierre for a side kick."

25. JODI ERICSON: VICE PRESIDENT

As the President's selection, Jodi breezed through the political convention in Houston without incident. A few feeble attempts were made to put Congressman Jesse Rowan on the ticket but the President's team railroaded the effort out the back door.

The following months were a blur of activity. Jodi had never worked such long hours and so hard in her life. The demands were beyond her wildest nightmares. This time her efforts weren't confined to the state of Nebraska or of Nevada. This time she was obliged to set foot in every part of the nation. One thing she had learned in campaigning, TV is great; but nothing beats the people seeing you as one of them and you seeing the people as individuals rather than globs of voters on a map.

Her schedule required coordination with the President. He took the high road of the big cities and critical states. Jodi took the low road of the byways and areas which had never met a President or Vice President. This she enjoyed. She was more at home with individuals who felt frustrated with big government. She remembered how her father and grandfather had been forced to fight government throughout her entire youth. The scars still burned.

Jodi refused to appear before the White House Press Corps. When she had a major speech to make, she chose a small town or city away from Washington, DC. At those sites she invariably gave half-hour interviews to the local TV stations. If media majors wished to participate they had to make their peace with the locals. Once it caught fire, this small tactic tickled the populace who enthusiastically supported the idea. Jodi's only exception was a promise to Marvin Pringle to appear on his show. He had earlier supported Jodi when senator.

At a rally in Missouri a bitter farmer wiped out by the floods interrupted Jodi's talk.

"What makes you think you can do anything about throwing out those crooks in Washington," he hollered belligerently.

Jodi looked down from her podium on the back of a truck. She had purchased a second-hand six-wheeler to carry all her campaign paraphernalia. She felt at home and so did her audiences. Her answer took the gathering by surprise.

"I can't, sir," she said, singling the man out and speaking directly to him. "You're 100 percent right. I can't do one damn thing about it."

"Then what the hell..." the man yelled.

Jodi interrupted. "You're the only one who can do anything. Get mad...get off your dead haunches and harness that congressional mule or your fields will never get plowed. That congressional mule will continue to eat what it pleases."

"They got my fields. They got my money. There's nothing else." The man was almost in tears.

"There's your blood."

Jodi was determined to brutally drive home the insensitivity of congress. Although she felt deep sympathy for the farmer perhaps it took deep tragedy--another Pearl Harbor--to awaken the American people to what they were facing: a relentless drive by government to extract every possible dime from the people.

Congress and the bureaucrats were certain they knew better than the individual how to spend his hard-earned dollars.

"The serfs had it easier, the Czars only wanted one-third," Jodi said, ignoring the TV cameras grinding in the background. "Americans now pay over half."

"We need another Boston Tea party," yelled a red-faced man in overalls.

"Perhaps we do," said Jodi. "American taxpayers have suffered 100 times the slings and arrows of the tea tax."

The next day Jodi was speaking from a high school stage in Arizona when a heckler stood up and protested.

"Congress takes our money for every fantasy they can dream up. My grandparents were German. They learned a new language. Now my taxes have been raised to support multi-lingualism. What's the matter with making today's immigrants learn English the way my people had to?"

"Nothing," said Jodi. "My grandfather came from Russia. He learned English without giving it another thought."

"They're forcing us to pay to keep alive cultural differences," the speaker cried angrily. "In our community gangs have formed along ethnic and language lines, which are kept alive in the schools."

"It's a problem across America," said Jodi in a calming voice. "Congress supported Departments of Education and Labor in funding for every language group in America. Language and cultural differences isolate communities."

"You agree English should be the primary language taught in all our schools?"

"Of course I agree," said Jodi. "Otherwise we face gradual disappearance as Americans. We easily can become another Bosnia."

Jodi waited until the applause, led by the questioner, died down.

"Today we as a people have no choice. Congress must be reigned in, controlled or replaced. Our survival as a nation in control of its own destiny is at stake. The President has a plan to harness the spendthrifts. I hope you will support him."

"What's his plan? We've heard promises before," hollered the original heckler.

Jodi studied the incensed skeptic who refused to sit down. Undoubtedly he'd been abused by an agency of government. She took a deep breath.

"You've got every right to be angry. Congressmen seem unable to get the message. Those who threw out the Democrats under President Clinton still compromised with the entrenched inmates."

A man in a red and blue plaid shirt yelled, "Make sense, lady."

Jodi said. "Let me cite an example. In school certain kids feel free to act up, raise hell and defy the teacher. They know the teacher can't discipline or expel them. So they cater to the bullies in the classroom. They know these bullies have real power and can hurt them during recess or after school. Yet we blame the teacher for everything which goes wrong, despite social workers rounding up more truant bullies to shove back into the classrooms. When it comes to protecting teachers, the NEA is useless garbage."

Jodi held up her hand as a man started to interrupt.

"Bear with me. The same thing happens to members of congress. They've let the bureaucracy become mad social working zealots who torment citizens. EPA, DEA, EEOC, ATF and hundreds of agencies make life miserable for our citizens. Hardly a single citizen today has escaped an ugly encounter with a bureaucrat. So the voter vents his spleen on members of congress because he can't reach the arrogant autocrat. Until our representatives wake up to the fact they've got to throttle such bullies, they'll continue to be maligned as incompetent toadies.

"Tell your congressmen, ***it's the bureaucracy, stupid."***

President Roland won reelection with 69 percent of the vote, sweeping Jodi Ericson along with him as the first woman Vice President of the United States.

26. HOME PORT

Jodi luxuriated by the pool with her husband...her first vacation since becoming Vice President.

"I'm proud as hell of your accomplishments," said Guy, "but they sure put a damper on my sex life."

"Mine too," said Jodi, sitting at Guy's feet and stretching her long legs into the water. She wore a two-piece light tan bikini which blended into the color of her skin. It took a second look to confirm the bathing suit. A solid figure of breath-taking proportions greeted Guy as he devoured her figure ceaselessly. She lifted her face to his, the blue-green eyes sparkling.

"Last night you explored every inch of me, now you're doing it all over again."

"This time with my eyes," said Guy. "Last night was with my body. I prefer last night's arrangement."

"Me too. But even whales come up for air."

"Watching you makes me think I've had enough air." He reached down and ran his hands through her hair as it spilled over his knee where she rested her head.

Married two full years Jodi still felt the electricity course through her body at the touch of Guy's fingers. From that very first time, in a semi-feverish state from the slight gunshot wound, when she had half-consciously allowed him to climb in bed with her, Jodi had entered another world. A world of unbelievable ecstasy. Far removed from her previous one filled with hate and bitterness. If there were such a thing as sexual slavery, she had willingly become one. Yet, like church and state, she had managed to keep politics out of the bedroom.

"You're unduly pensive," Guy said, running a hand down her neck and along her shoulders. "Something wrong?"

"No. Only concerned. I'm so much in love and soaring so high I'm afraid I'm like Icarus...flying so close to the sun a crash lies ahead."

"Nonsense. Superstition. You've done wonders."

"Mostly with your backing?"

"And blessing, don't forget that."

"You know I never will."

Guy chuckled. "I've no doubt. You never forget anything, especially the time I refused to give you a job. Yet see how destiny steered you away from me at the right time and threw you into my lap when it counted."

"You believe in destiny?"

"Yes," he said. "If you take into consideration luck, birth, drive, timing, environment and a multitude of other factors."

"I think for the most part we shape our own destiny."

"You suffered a few setbacks; yet you maintained full speed ahead," smiled Guy.

"I knew you would always be there."

"Nonsense. I'm talking about the days before we met. You were a pretty independent duckling."

"An ugly duckling?" Jodi teased.

"A damn beautiful one," said Guy. "That's what originally made me fight you."

"Fight a good looking girl? You've got different genes than the rest of mankind?"

"You know better," Guy gave her hair a slight yank with a light laugh. "Beautiful gals dominate Vegas. That's why I was suspicious. They all want something."

"I want something."

"Yeah?"

"Your body. As long and often as I can possess it."

Guy grinned. "Well, young lady. That's a two-way street." He stood up. "My bathing trunks can't handle the merchandise any longer."

Jodi glanced around the hotel pool before rising and slipping a hand in the back of his waistband, tugging hard at the brief trunks.

"There. That should tighten the desire," she whispered. "We'd better hurry before you explode in front of all these people."

* * *

Late that evening, while Guy was working long distance calls to MRI operatives in the Ukraine, Jodi decided to sneak out and stretch her limbs.

Dressed as a sightseer in slacks, tennis shoes, and her hair tucked under a baseball cap, she leisurely strolled the strip. She walked about unrecognized, although it probably would have made little difference since famous people were commonplace in Vegas.

A boy of about fourteen stepped in front of Jodi.

"Got a buck for a sandwich, mister?"

On closer inspection, he corrected himself. "I mean, lady. Could sure use some grub."

The youth's demeanor was brusque, but not threatening. Jodi determined the kid truly was hungry. She hesitated only momentarily.

"You hungry? Or want money for a quick fix?"

"You deaf? Said I was hungry. Don't rap drugs."

Jodi glanced at the neon lights blazing two doors away. McDonalds. She made a sudden decision.

"Never hand over money. But am a bit starved myself. So if you'd care to join me, it's my treat."

Torn between suspicion and an empty stomach, the empty stomach won out.

"Okay," he said with sullen reluctance. "You're not a truant snitch?"

Seated across a table, Jodi watched the youth gulp down french fries, a double-decked hamburger and a giant sized malt. She had a coke and nibbled on some french fries.

"You really were hungry?"

The youth looked up with a disgusted look but kept on eating. When he had finished his malt he sat back and grinned.

"You work one of the tables? "

"No. What makes you ask?"

"Workers only suckers who give handouts. Big-time gamblers kick us off the streets."

"Why aren't you in school?"

"What for? Two older brothers finished school and they out of work same as me. Don't teach you nothing 'bout getting a job." He looked Jodi straight in the eyes for the first time. "If I'm gonna sit here and answer questions, I could use another shake."

Jodi smiled and slipped him a dollar bill.

He grabbed the bill and looked down on her as he stood up.

"Ain't you scared I'll run off 'n buy drugs?"

"Druggies don't eat like you. I'll take that chance."

Back at the booth the youth studied Jodi unabashedly.

"Bet they hired you cause of your looks. You ain't savy enough to run a table in this town."

"You think I'm a softy?"

"Yeah."

"Ever arm wrestle?"

"Do it all the time. Told you I had older brothers. Why?"

"Thought we might try one game."

The youth looked at Jodi like she was nuts.

"You gotta be kiddin'. Arm wrestle a dame? I'd wreck your arm."

Jodi said. "Slide your malt to one side and let's have at it."

Skeptically. Thinking at first he was being led into some kind of trick, the youth grinned and placed his right elbow in position. Two minutes later he rubbed his arm muscle in defeat, staring at Jodi in disbelief.

"You're some kind of weirdo. Good looker. Running around dressed like a man. Where'd you learn to arm wrestle?"

"On a farm in Nebraska. Now. Give me an honest answer. Why aren't you in school?"

"You don't listen good. Teachers spend all their time arguing with bullies or calling on teacher's pets. Rest of kids don't matter. Just like home. Older brothers run the place. No father. My old lady works as rest room supervisor in casino."

Jodi sighed. It all seemed so hopeless. She remembered her father's quote from Art Buck.

"How soon do we forget
 What elders used to know:
That children should be raised,
 Not left like weeds to grow."

27. JODI TACKLES TV MONSTER

Keeping her word to Marvin Pringle to appear on his Sunday TV show, Jodi Ericson reached the studio one-half hour early on the arm of Tony Pellini. Only then did she learn the other guests were Speaker Metzer, "Jam" Johnson--the regular talking head, and Congressman Reverend Jesse Rowan, Chairman of the House Judiciary Committee.

"Out to tear your guts out," fumed an incensed Tony. "You ought to withdraw. They had no right to invite only enemies."

Senator May Trudeau disengaged herself from a group of admirers and came over.

"Looks like they stacked the deck against you, Jodi," she said in her low lisping voice. "Had I known I would have warned you." Senator Trudeau wore a short navy blue skirt with an open yellow blouse and blue shoes with spiked heels.

Senator Trudeau made no apology for her flamboyant dress. She flouted her feminism and disdained slacks, suits and any form of aping men's attire.

"I'll survive," Jodi said, tossing her cascading black hair away from her face. "I'm only facing domesticated cows, not raging wild bulls."

"I'm not so sure," said Tony. "It's the ambush, not the frontal attack, you've got to guard against. Speaker Metzer shakes your hand with a cobra up his sleeve."

Senator Trudeau laughed. "One of these days that cobra will bite it's keeper."

"Then there's "Jam" Johnson," said Tony.

"A comic relief," said Jodi. "He's got the mental arm of a midget and his fast balls barely travel thirty miles an hour. Main problem with him, he's belligerent and loud. Hardly dangerous."

"What about Reverend Jesse Rowan?" Tony glanced at this wrist watch. Ten minutes from show time. "What's his ax?"

Jodi confessed what the President told her.

"He's full of hate because President Roland slipped me in as the new face instead of him. The Speaker and black caucus tried to promote him for VP at the convention but got nowhere."

Senator Trudeau worriedly fingered the silver earring on her left ear. She spoke in an undertone, glancing at the talkative crowd taking seats in the studio theater.

"Congressman Rowan is dangerous. He's a non-stop yak artist with a megaphone. And he's vindictive."

"Why, for God's sake?" Jodi was upset.

"He takes his politics personal," said Senator Trudeau. "Any criticism is slander. As head of the black caucus, he finds hidden motives in his bean soup...motives he must expose and rail against."

"Same old theme," said Tony. "My ancestors were mistreated, so I've got to mistreat you." He snorted in disgust. "My grandfather was killed when Italy was drafted to fight Hitler's war, but I don't run around with a megaphone decrying the Germans."

"Pays off politically."

"If the truth were known," Tony said in sour tones, "none of us really ought to celebrate our roots. Romans captured more slaves than U.S. merchants ever did from Africa. The English purposefully starved millions of Irish who they thought were multiplying too fast and replaced them with 800,000 cattle. We know what the Germans did to the Jews. Russia, under Stalin, eliminated hundreds of thousands. The supposedly peaceful Incas conducted hideous pagan slaughters and the Spanish Inquisition was an abomination. The numerous crusades exterminated entire populations justified by ideology."

Jodi interrupted slyly.

"Appears only Genghis Khan conquered nations and established governments without regard to race, creed or color."

Tony admitted Jodi's point with a skeptical grin.

"Understand his dynasties lasted because they tolerated the divergent views of the conquered people although they came down hard on law and order?"

"Like Justice Holmes said. *Give people a choice and they'll opt for order above all else.*'"

"Reverend Rowan's no Justice Holmes," grimaced Tony. "He thrives on anarchy. Like Robespierre, he gets the attention of the lawless."

"Congressman Rowan's got a scary personal vendetta," said Senator Trudeau with an apprehensive glance at Jodi.

Tony caught the concern. "You talking about the convention put down by President Roland?"

"Exactly," said Senator Trudeau. "Rowan is enraged. He feels Jodi cheated him out of the Vice Presidency."

"The President's decision, not Jodi's," said Tony.

"Makes no difference. I told you the man takes everything personally," said Senator Trudeau. "Now he's swung the black caucus against the President and Jodi...a hate war, no matter the program."

"All were nominally on the President's team until he used Jodi to shut them out at the convention," said Tony.

* * *

Moderator Marvin Pringle introduced the guests.

"Fat Albert" Metzer spoke from the Speaker's office in the U.S. Capitol. He felt it set him a bit above the others.

Moderator Pringle led with a question.

"The news this past week has focused on one more attempt to get fiscal control of government. We'll turn to Speaker Metzer first. Can congress find the answers?"

"Only with gradual 10 percent cutbacks," said the Speaker. "Something congress can live with."

Pringle turned to Congressman Rowan. "You agree?"

Reverend Jesse Rowan, a rather handsome man with light skin and a carefully trimmed pencil mustache, loved the sound of his own voice. The Elders once advised him to limit his sermons to thirty minutes because parishioners were slipping out before the collection plate was passed.

"The Speaker has the right idea," boomed Reverend Rowan. "Start with 10 percent. No more. It's the people at the bottom of the ladder who get squeezed. Civil servants shouldn't suffer because cockeyed schemes proposed by elite snobs promise to reduce costs. A bureaucracy is the employment of people. Ten percent off the top of each agency would be plenty. Cut more defense..."

Pringle didn't wish another long-winded diatribe against the military on his TV show. That problem had been hashed to death. He interrupted.

"We'll now hear from Vice President Jodi Ericson. Do you think ten percent off the top is the way to go?"

Jodi glanced at the front row where Senator Trudeau and Tony Pellini were sitting to give her encouragement. Tony gave her a quick thumbs up signal.

Jodi turned to Marvin Pringle.

"I think the whole idea stinks," she said with a smile. "Ten percent is like the old story of the man with ten mistresses. He decides to cut costs. So he cuts out breakfast and lunch to each. Soon he ends up with ten sick mistresses in the hospital."

"You're solution?"

"Get rid of a few mistresses. It's that simple. Congress never asks whether the activities of certain agencies are really necessary? Aren't ten mistresses a bit much? Congress has thousands."

Pringle laughed, finally forcing a straight face.

"In my house, bringing home even one mistress would get an ice pick in my back."

Said Jodi. "We're on the same silly no-result track Clinton and Gore pursued. They achieved no reduction in costs. Instead, a painful system of retirements, attrition, retraining, domino-bumping and job-shifting took place--but precious few dollars saved. To improve government, we've got to get rid of big chunks. Percentage bites are mysteriously replaced. Total agencies need to go."

"You don't have much compassion for the little fellow, do you?" exploded Congressman Rowan.

"I certainly do," retorted Jodi. "Never occurs to you that compassion may be due the taxpayers. As Sowell argues, in the media 'greed' refers only to people trying to keep what they've earned. Especially the little fellow forking over money to support bureaucrats earning three times as much as he does."

"You realize under your budget cuts blacks would be the first to go?"

"Nonsense. The exact opposite. Your 10 percent solution would oust far more blacks. Particularly those at the bottom of the ladder," said Jodi. "I know of no total black agency. Most are at least 80 percent non-black. Blacks would fare best when an entire agency is terminated."

"Looking back, President Clinton and Vice President Gore's efforts were the most compassionate," snapped Rowan. "For the first time federal employees were the principle source of ideas."

"You'd reform the penal system by putting prisoners in charge?" Jodi Ericson gave Reverend Rowan a scathing look of disgust. "Next, I suppose you'd have serial rapists and killers set prison terms? That proposal was a Clinton-Gore sham to appease the heavy bureaucratic vote. Like you, Gore never saw a tax he didn't vote for."

Reverend Rowan ignored the remark.

"Another positive was Gore's promise to 'treat taxpayers like customers.' We should adopt that idea today."

"Another fraud. Customers in the market place choose to spend only what money they feel they can afford. They return faulty items and get their money back. Gore lied. Where was the taxpayer's choice...the chance to get his money back?"

Congressman Rowan stared at Jodi, his face flushed with anger.

"You and the President are determined to fight any solutions proposed by congress?"

"Not solutions, only more layers of bureaucracy called solutions," said Jodi, smoothing her cashmere skirt and crossing her eye-catching legs. She smiled at Tony and Senator Trudeau.

"You despise government workers?" Congressman Rowan pressed hard on his mustache, biting off his words.

"What a silly accusation," said Jodi. "I don't despise children in an overcrowded schoolroom. I want them placed elsewhere. Let them learn, contribute, become productive and prosper. I want our bureaucracy weaned from sucking up taxes and put out into the real world where they can learn a trade besides writing and rewriting regulations."

Congressman Rowan moved belligerently to the front of his chair.

"You forget the services bureaucrats perform."

Jodi retorted. "Producing regulations which kill jobs and drive up costs that stymie economic growth? Services like sending home-owners to jail for filling in mosquito beds in their back yard? Or printing more tax forms?"

"Hokum," challenged Rowan.

"Is it?" Jodi asked, glancing at Moderator Pringle, expecting him to get Congressman Rowan off her back. "One-fourth of every health dollar is spent on regulatory and administrative costs. Worse, 95 percent of the cost of children's vaccines are due to regulations."

Marvin Pringle finally stepped into the middle of the heated exchanges before they got completely out of hand.

"Let me bring in our own regular commentator, "Jam" Johnson. He's been traveling across the country this past week."

"Where I've been," Johnson said, annoyed that he'd been excluded from center stage for so long, "people are primarily concerned about schools and crime."

"My reading precisely," interjected Speaker Metzer. "We hear too much yapping about taxes."

"We've got to put more money into our schools and combating crime," said Johnson.

"Another moronic solution," said Jodi, realizing her harsh response was throwing down the gauntlet. But the time had long since passed for bland compromising apologies for every problem.

"You think educators don't need it?" challenged "Jam" Johnson.

"I think they've had enough money," said Jodi. "Now it's time they learned how to use it."

The Speaker leaned forward into his desk microphone and said with rasping sarcasm.

"You know all the answers? How to run the schools? How to handle crime? I congratulate you on your omniscience."

"A child with a dime knows better than to order a pound of chocolates he can't pay for," said Jodi peevishly.

"Just what's your solution? If you have one?" Congressman Rowan demanded. He was certain President Roland had picked Jodi as his running mate because of her cover-girl ability to get media attention. He had far more right to the job than any woman.

Jodi answered slowly and emphatically.

"Schools have lost their way. Violent and disruptive students make education impossible. Academic excellence demands their expulsion."

"Jam" Johnson could barely hide his elation. He sensed he had found Jodi's Achilles heel.

"Throwing kids into the streets to commit more crime?"

"If schools can't introduce discipline," said Jodi, "both whites and minorities will be utterly useless in the coming high-tech world of the twenty-first century."

"Communities are responding to minority challenges," declared Johnson. "I interviewed hundreds in depressed minority areas."

"Not so," said Jodi. "In Florida, for example, minority zealots demanded English be replaced in all textbooks, libraries, business documents and even road signs."

"What's wrong with people demanding their native languages in the schools?" Congressman Rowan challenged. "They want to hang on to their roots."

"Fine. But not in school," said Jodi. "If dinosaurs moved into Florida I suppose it would be proper for those same leaders to make all citizens learn dinosaur language."

"How ridiculous."

"By way of contrast, in Singapore 3 and 5-year olds learn English--the language of computers, international business and commerce," said Jodi. "Our friends in Florida could better spend those bilingual education millions on computers.

"What about the kids thrown out into the streets?" demanded the Speaker. "They will be free to commit more crimes."

Jodi took a deep breath. This is something she and President Roland had discussed at length. They had decided only one long-run solution was possible.

"Young men in prison cost better than $25,000 per year. When released back into society they're usually far worse than when they entered." Jodi fell quiet, gathering her thoughts.

"We all know the figures," said "Jam" Johnson. "We're listening for solutions."

"I propose a Youth Work Corps. YWC. Disciplined learning camps for all youths not in school. Run by the military, the camps would provide education and training with no stigma attached," said Jodi. "They'd be patterned after President Roosevelt's hugely successful CCC camps."

"Our laws won't permit such idiocy," declared Congressman Rowan.

"New crime waves are by children under age 18, not adults. We need to get control of delinquents 12 to 17 before they become hardened criminals."

"Such kids can't be controlled."

"General Hardin would operate the camps under the Universal Code of Military Justice," said Jodi.

"Illegal," shouted Rowan.

"A matter for the courts to decide at some future date."

"The media will decide long before the courts," cried 'Jam' Johnson. "We're the watch dog of society. We'll keep such a harebrained scheme from ever happening."

Jodi hesitated. She knew she was treading on hallowed ground. The media had captured the First Amendment as if it applied to them only. Like judges in the Spanish Inquisition: they thought their way was the true road.

"The public is beginning to look upon the media," Jodi declared resolutely, "not as a friend and safeguard, but as another con-artist hawking its own Ponzi schemes."

"An insane remark," shouted Johnson.

Jodi smiled. Nothing proved a point better than an opponent's rage.

"Shakespeare suggested 'kill all lawyers.' A bit drastic but perhaps a national board should set fees for lawyers as congress proposed for doctors and dentists. In that regard I've seen no media fighting to help the taxpayer in curbing exorbitant malpractice costs. Perhaps we've overindulged both the media and the legal system with the first amendment."

"The media's exposed malpractice and insurance costs," retorted Johnson. "It's up to congress to take action."

"In some rural areas not a single doctor will deliver babies because of insurance costs," said Jodi. "I've seen no media crusade on this."

"Congress dances to the tune of American Trial Lawyers, not the media," said Johnson.

Jodi lifted her face to Johnson and spoke quietly.

"The media covers a supreme court nominee hearing like a wet blanket. But it's woefully silent when it comes to protecting the public against lawyers--including those sitting on the bench."

"You suggest the media go witch hunting?"

"No. Fact finding. Reporters want to produce another Watergate to give them fame and fortune. Well, they've got one in our lawyer-controlled judicial legislators. EPA decisions alone would shame Watergate, the Teapot Dome and Boss Tweed."

"Judges are appointed," protested a confused Johnson, shaking his head. "We can't touch them."

"To survive, the media needs democracy as much as the people," said Jodi. "So you'd best find a way to dethrone those who abuse their power to reshape society to suit their elite whims."

Johnson's comment was sarcastic.

"Repudiate the constitution?"

"The next despot to subject the people will not be czars or dictators, but the spider web of lawyers, judges and poisonous regulations cast over the nation."

Rowan retorted. "Many presidents have been lawyers."

"The best presidents have been ***non***-lawyers," said Jodi. "Starting with Washington and ending with Roosevelt, Truman, Eisenhower and Reagan. Abe Lincoln largely educated himself without benefit of prestigious law schools or Oxford scholar robots."

"Wrong again," Rowan corrected eagerly. "Franklin Roosevelt was a lawyer."

"Not so." Jodi smiled disarmingly. "Roosevelt attended, but never finished, law school."

"Makes no matter," complained Rowan. "Lawyers enacted our laws."

"To the tune of exhorbitant fees to their profession. Malpractice suits, innumerable class action litigation--from which lawyers end up with more money than the damaged victims..." Jodi caught her breath.

"Out of twenty-eight billion dollars, that's ***billions***, spent on atomic waste, only ten percent actually cleaned up any waste. Additional millions were paid lawyers by victims targeted by EPA,

DEA, EEOC, OSHA, and numerous other rapacious agencies created by congress to trap innocent taxpayers. It's time to stop the orgy of federal persecution."

"Nonsense," sputtered Rowan.

"Congress has created statutes calling for penalties up to $25,000 per day in civil cases, and months and years in federal prison for criminal cases, without providing proof of any environmental damage or intention to violate a single provision in thousands of pages of federal regulations.

"King George was a Mother Teressa compared to some agency zealots," cried Jodi.

"Our courts prevent such excesses," retorted Reverend Rowan.

"In a pig's eye," declared an incensed Jodi. "There's no greater indictment of our judicial system than the fact that honest men are afraid to go into court, while criminals saunter out revolving doors.

"Einstein was on the money," Jodi said, "when he declared 'bureaucracy is the death of any achievement."

"And just how would you overcome all your imagined problems?"

"Support my amendment for term limits for all judges; publicize hundreds of arrogant case histories; let the people know the magnitude of the freedoms judges are taking away. The media owes such an investigation to its own future salvation."

28. GUY ERICSON AND PRESIDENT

"Watched you on TV," Guy Ericson said to his wife over the telephone. "You were spectacular. Sorry I couldn't be there but you had Tony for support."

"Missed you too," said Jodi. "Sheets get cold and clammy with only a single body to heat them."

"I've lain awake many a night wondering about priorities. My business success or your political success." Guy paused as he sipped on a drink. He sat beside the hotel pool in Vegas at four o'clock in the afternoon. Jodi was still at her desk as dusk settled in Washington.

"We wanted both."

"I know. Eager beavers. Now I'm beginning to wonder if it isn't more important for you to be here at home?"

"Home's wherever you have me in your arms," Jodi said, her voice husky. "I'll be there this weekend."

"Don't bother. I won't be here."

"What you talking about?" Jodi cried, sitting up in her chair and clutching the phone until her knuckles stretched taut.

"The President just called," Guy hastened to explain. "In the morning I fly to New York and meet with top scientists. Then on to London to rendezvous with the President for a trip to Russia and the Ukraine to check on the atomic weaponry. Remember my company, Material Resources International, is the prime contractor and the U.S. has a lot of money sunk in the dismantling process?"

"Oh," said Jodi weakly. Fate seemed to throw her one curve after another. A sudden thought worried her. "You won't be in any danger?"

"Naw. I'll be with the President. He'll be surrounded by more security than Ft. Knox." After a moment, he added. "By the way, the President saw your performance. Said you were terrific...that I was one lucky fellow."

Jodi felt she would be a lot luckier if Guy visited her instead of New York City.

"Can't you possibly swing by here? If only overnight."

"You know I'd love to, Jodi. My schedule's tighter than a rubber band stretched to the breaking point."

Jodi was silent as she glared at the telephone, as if somehow it was at fault.

Guy's voice dropped into soft conciliatory tones. "Tried to get the President to let you meet us in London. He replied with a firm no. Said the President and Vice President shouldn't be out of the country at the same time."

"I understand. It's a long standing rule."

"There's nothing wrong with meeting you in New York on my return trip," said Guy cheerfully. "Book us into the Plaza. I'll call from Europe. Okay?"

"You know damn well it is," said Jodi, attempting a smile. "Stay away from night life in Paris."

Guy laughed. "You've got no worries. We're flying over France directly into the Ukraine. Take care. I love you."

After he hung up Jodi sat staring at the telephone. Slowly she replaced the instrument in the cradle. She wondered how that Viking bastard had so completely crawled inside her very soul. In her youth she had dismissed the best looking boys with the snap of her fingers. Now she had become an emotional slave.

It didn't speak well for the independence of her feminine mystique.

*   *   *

Less than a week later Jodi awakened cursing at the simultaneous ringing of her telephone and the doorbell to her apartment. She had preferred a small apartment instead of the cavernous Naval Observatory since she considered her real home with Guy in Vegas. Many Vice Presidents had lived in apartments. Furious, she threw on her purple robe, picked up the telephone and hollered.

"Hold a damn minute. Someone's ringing the doorbell and banging on my door." She lifted the security peephole. "What the hell's going on?"

"It's Keefe O'Hara of the FBI," said a red-faced Irishman, holding up an identification badge. "Please open up. We need to talk."

Jodi's face blanched. Instinctively she knew something was dreadfully wrong. She opened the two locks and removed the chain, admitting Inspector O'Hara. She closed the door behind him, shielding out a hall full of bodies.

Inspector O'Hara was obviously ill at ease. He toyed with his earlobe.

"Please sit down, Mrs. Ericson," he said, escorting her to a high wing-backed chair. "I've some dreadful news."

"Something's happened to my husband?" Jodi said. Her heart pounded and she felt her lungs nearly collapse from the suffocating pressure.

Keefe O'Hara wasted no time on softening the blow.

"Your husband's been murdered in London..."

"Oh no...oh, my God no." Jodi barely managed to escape fainting, grabbing the arms of the chair for support.

"Along with the President of the United States," O'Hara continued in a monotone.

Jodi looked at O'Hara with eyes full of pain, unbelieving.

"Guy and the President?"

Jodi sat for a full minute trying to absorb the tragedy. Tears refused to come as she stared into space. Finally, without a word, she got up and went into the bathroom, closing the door behind her. For a full fifteen minutes Inspector O'Hara waited patiently. At last Jodi returned, her face scrubbed and her hair combed.

"I must get dressed," she said mechanically, retreating to the bedroom.

Another half-hour passed before Jodi emerged in a black skirt and jacket with an open collar white shirt. To Inspector O'Hara she seemed quietly composed and aloof.

"How'd it happen?"

"Two Shiite assassins posing as hotel employees managed to infiltrate through the hotel's freight elevators. All personnel were checked for weapons but these two had managed to previously stash hand guns inside the Presidential suite's linens."

"Anyone else killed?"

"No. Two secret service wounded, but the assassins had their targets picked out."

"Capture the assassins?"

"One killed instantly. The other died in the hospital forty minutes later, conscious and babbling apologies to Allah. They expected to die. Papers commended their bodies to Allah."

"Guy thought all exposures were behind them," said Jodi. "I talked to him last evening. He'd just returned from danger zones in Russia and the Ukraine where security had been intense."

"Scotland Yard and our own security thought they had all terrorist avenues blocked," said Inspector O'Hara, his eyes focused on the rug. "The assassins had false hotel employee identification tags."

Jodi said through gritted teeth.

"Guy and the President worried often about the implacable hatred of Hezbollah foes for blocking the sale of weapons to Iran."

"The assassins weren't after the President," said O'Hara.

Jodi's voice broke.

"My husband? Why him?"

"Not him, either," said O'Hara hesitantly, dropping his eyes to the floor. "They were after you."

Jodi's face blanched. "Me? How ridiculous. I'm nobody. Where'd you get such a cockeyed idea?"

"From the assassin who died," said O'Hara. "He kept pleading with Allah to forgive him. He'd made a mistake."

"A mistake?" Jodi echoed, stunned to the point of shock and trying desperately to make sense of O'Hara's conversation.

Said the FBI chief.

"Scotland Yard operatives who rode in the ambulance picked up the information from the wounded assassin's sometimes incoherent ranting," said O'Hara. "The assassins were told to eliminate a woman who often wore purple traveling with her husband and the President."

Jodi thought a moment.

"My office mistakenly issued an itinerary which had me meeting Guy and the President in London. But it was retracted when the President nixed the idea."

"No doubt in small print on a back page," said O'Hara.

"I don't remember seeing it. I wasn't paying attention."

"Anyhow, your husband appeared in a purple bathrobe in the doorway with the President just across the hall from his own room. They were emerging from a late night session."

Jodi sobbed. "Guy had a bathrobe matching mine. A Mr. and Mrs. arrangement."

O'Hara explained. "On seeing the purple robe the assassins opened fire immediately. They were nervous amateurs. Only when they saw the men fall did they realize their mistake."

"My God," cried Jodi. "Makes me responsible for my husband's death."

"In no way," the FBI chief said firmly. "Assassins are irresponsible scum. They've absolutely no understanding of the value of human life."

Tears welled into Jodi's eyes.

"Poor Guy. He bought that bathrobe to please me. Purple was dad's favorite. A nostalgic hangover from his ancestor's coat-of-arms, in which purple was the dominate color."

"Better throw the garment away," said O'Hara. "The assassins aren't through yet."

"What do you mean?"

"They didn't achieve their objective--you. So the culprit behind the murders is still at large. From now on your job is to be doubly alert.

Jodi glanced at her wrist watch. Four-thirty in the morning. She wiped her eyes with a large white handkerchief.

"Heathens," Jodi muttered under her breath. Aloud, she said.

"What's our next move?"

"Soon as you're ready, Mrs. Vice President," said O'Hara. "The Chief Justice is waiting in the Capitol to administer you the oath of office. Since your husband was involved, the Chief Justice wanted to wait away from your apartment until you felt able to join him."

"I'm able," said a grim Jodi, wishing she had never come to Washington in the first place. But that was a bit late now.

29. PRESIDENT JODI ERICSON

Weeks of sleepless nights remembering Guy plus violent budget battles with congress drove Jodi to the point of exhaustion. Congressional greed sprung up like a garden of mushrooms upon Jodi's assumption of the Presidency. Speaker Metzer and his cohorts were determined to set up the new female President as a patsy for their many pork programs. President Roland's death shattered what little party loyalty remained.

Special interests seethed with unrest to bring back programs kept in check by President Roland after the austerity introduced by the Newt Gingrich and Bob Dole congress. A weak woman president was their golden opportunity.

Congress, choking at the bit to take over once again, hurriedly pushed five new taxing bills for pet projects up to the White House for signature. Jodi vetoed each one, managing to sustain the veto by only three votes on the most critical spending measure. Congressional leaders, the black caucus, special interest groups demanding the replacement of social programs were furious. They had not expected such iron-willed resistance from a woman.

This morning they were meeting with President Jodi Ericson in the Oval Office. Speaker Metzer, Senate Majority Leader Meyer, Black Caucus Leader Rowan, and two other members of the House were present.

"Madame President, your vetoes are killing our promises to the people," said an incensed Speaker.

Congressman Rowan fought to keep a civil tongue. He remembered his unhappy confrontation with Jodi on public TV.

"Speaking for the Black Caucus, we consider your refusal to provide more help for welfare, food stamps and public housing a direct slap against the black community."

"Where am I supposed to get this money?" President Jodi Ericson inquired gently. "You're already extracting $26 billion each year from the taxpayers for food stamps alone?"

"From new tax bills you rejected so out of hand," prompted the Speaker.

"Presidents Reagan, Bush and Clinton all succumbed to enormous tax increases--each one bigger than the preceding one," said Jodi. "Clinton's was the largest in the history of this country. Newt Gingrich's efforts only slowed appetites. The bloated bureaucracy wants still more. Yet no problems have been solved."

"The problems are huge..." said Congressman Rowan.

"So is the spending."

"Our new tax bill will finally solve the last lingering hurdles," said the Speaker. "I've mustered majority support from each house."

"I've studied your new tax bill. Another budget buster which pushes our people into complete socialism--barely one step ahead of Communism which a few short years ago we railed against as an evil empire."

"This new bill should be the last of the tax hikes," said Senator Meyer, studiously biting a rough nail. He refused to look up into Jodi's angry eyes.

"The nation's heard that old song before. President Reagan believed congress when they promised three dollars in cuts for every dollar in taxes," fumed Jodi, rising from behind the desk and pacing before the open windows looking out into the Rose Garden.

"Congress lied. They had no intention of keeping their part of the bargain. President Bush said 'read my lips, no new taxes.'...a promise congress turned into another lie which led to his defeat. Next President Clinton came up with a third big lie, duping the voter into believing he was only going to 'tax the rich.' Instead he and congress created the biggest tax burden in the two-hundred year history of the republic. Finally a Republican congress failed its valiant effort to strip the bureaucracy down to a lean servant of the people. It's still a bloated hog. Goebbels told less whopping lies than congress. It's time the lies stop."

"Money is necessary to run a government of over a quarter billion people," the Speaker retorted. He had difficulty suppressing his undying hatred of Jodi. And now she was President. What a calamitous turn of affairs. Well, he had the muscle and means to make her Presidency the laughing stock of the nation.

Congressman Jesse Rowan's tone was nasty.

"Either be rich like yourself or starve? That your credo, Madame President?"

Jodi, spinning on her heel and facing the group sitting before her, minced no words.

"The poor soul who struggles to earn, save and own a home is a fool to believe congress--any congress--will lift a single straw off his back. You people are turning the American taxpayer into an overburdened jackass. You shoulder him with bigger and bigger loads until he finally breaks. Then you'll have your vaunted social state--exactly like the collapsed Soviet Union."

"Madame President, you're reacting hysterically," said Congressman Rowan. "Again, you forget the poor."

"On the contrary. I see human beings as more than domesticated animals on a permanent plantation of dependency. I see them capable of accepting responsibility, work and becoming the equal of any other human."

"In a democracy the poor have rights," protested Rowan, "and taxes support those rights."

Jodi returned to a standing position behind her chair. She directed her remarks straight at the Speaker.

"I intend to veto every single tax bill. Evidently you've no interest in cutting spending so I warn you now, you'd better start thinking about pulling fat tapeworms out of pet pork projects."

"You forget," said Speaker Metzer, "the power of the purse resides in congress. You were catapulted into office by a fluke assassination. The people are behind us, not you, Madame President. Private polls of members in both Houses assure me of the power to override. So don't hang yourself."

Jodi pressed the buzzer on the corner of her desk.

"In that case, Speaker Metzer, I assume this meeting is over. If you insist on bulldozing that tax beast through congress, I promise a veto."

Three weeks later President Jodi Ericson vetoed the largest tax hike in the history of the nation. The veto was upheld by a single vote in the Senate after Senator May Trudeau got her many supporters to back the President. The Speaker and Majority Leader were rabid. Their very leadership depended on such diverse goodies.

30. RUMORS OF IMPEACHMENT

President Jodi Ericson first read about the threat of impeachment in *The Washington Post*. Congressman Jesse Rowan, Chairman of the House Judiciary Committee, and Speaker Metzer had garnered the black caucus and a dissident group of angry special interest members of the House to draw up a bill of particulars. Contents of the bill weren't specified, only that members of the House, indignant with President Jodi Ericson's constant repudiation of every congressional initiative, had coalesced behind a movement to impeach.

Tony Pellini sat across the breakfast table as she finished reading the devastating news. He had rushed over with the paper before the media could get to her with any brutal questioning.

Jodi stared for a long time at the gilded clock decorated with an American eagle and a standing figure of George Washington. The eagle plus the figure of George Washington seemed to signify permanence and courage--qualities she needed at the moment.

"So the Speaker and his rabble-rousing buddy want to impeach me?" Jodi sprang to her feet with such force she sent her chair catapulting over backwards. Tony hastened to set it upright.

"They're firing blanks...noisy but no more deadly than firecrackers," said Tony, trying to calm Jodi's fury.

"Well I'm damned if I'm going to sit quietly awaiting the guillotine. Nixon sat on his thumbs and tried to carry on as if nothing was happening. Boris Yeltsin said 'so you want a fight?' He took his case to the people. Demanded a referendum and won. Later he had to dismiss an obstructive parliament, just like our congress. But whether he won or lost, he gave it his best shot."

"Exactly what leaders are supposed to do," agreed Tony. "Give it their best shot."

"For two decades our Presidents sat back and cringed before an Imperial Congress," said Jodi, striding back and forth behind the breakfast table. "The buck has stopped with this intended wimp."

"Always doubted the Senate could have garnered a two-thirds vote against Nixon had he hung tough," said Tony.

"In the beginning he'd only needed to admit an error in judgment."

"But he lied?"

"Right. And the people are sick of lies," said Jodi. "Foolishly he lied to cover up a burglary no more important to the nation than a gnat on the ass of an elephant. Yet Congress told three tax lies which can utterly destroy our form of government. Still it prances merrily along ready to commit more lies."

"Nixon foolishly played a cat and mouse game trying to hide the truth."

"An always fatal tactic."

"Delay or indecision is deadly in politics or war," said Tony. He watched Jodi calm down, resume her seat and sip her coffee.

"Nixon made a horrible mistake," Jodi said, talking as much to herself as to Tony. "If he'd been honest and fought back he'd never been convicted by the Senate. His buddies on the hill convinced him to resign and spare the country. They meant 'spare the congress.' Well, that's exactly what I don't intend to do."

"Spare the congress?"

"The country would have been better off if congress hadn't been able to enact a single law for two years. Equals sending them home for a full term."

Tony chuckled. "What a relief. A real gift to the nation."

"I want to confront every single hypocrite in congress. Hang him out to dry with me. Nixon tried to ignore the harpoons they buried in his back."

"What's your plan? How can I help?"

"I need to give it some careful thought," said Jodi, clasping her hands behind her head and snapping back her elbows to relax her shoulder muscles. Her breasts jumped tight against the soft cashmere sweater as she unconsciously took a deep breath. Tony again realized Jodi was a beautiful vibrant young woman.

"Those articles of amendment you spent so many hours going over with President Roland wouldn't be a bad place to start," said Tony. "They were heavily researched and designed strictly for the welfare of the people."

"My thoughts exactly," cried Jodi enthusiastically. She reached over and grabbed Tony's hand. "Two minds in complete harmony. I've been thinking of those articles ever since I became President."

"One big positive," said Tony, "Those articles rise above party politics. They should get people support."

"Also vicious congressional opposition."

"Like you said, give it your best shot. Give voters the facts and make them choose. Get off their potato couches. After 200 years, getting rid of another fossilized Continental Congress might galvanize the people into action."

"I'll do it," exclaimed Jodi. "A Kamikaze mission. We either succeed or get shot down in flames."

"I'd rather call it an attack on Hiroshima," smiled Tony. "Kamikaze runs didn't fare too well."

31. AUGUST: THE AMENDMENT

President Jodi Ericson canceled all appointments and went into semi-hibernation at Camp David for one week as she planned strategy. She decided on a daring approach. Remembering Yeltsin's early years, she realized leaving congress a platform to fire from was inviting failure. It was late August, six days until congress adjourned for vacation.

Like the duck on the Japanese pond, the following week Jodi conducted relations with congress smoothly and with tranquil courtesy...but underneath 'feet go like hell.' She got busy.

Jodi secretly summoned all those who had worked on President Roland's constitutional amendment and sharpened the twelve articles with revisions of her own. Covertly she called Marvin Pringle and made arrangements to appear on national TV the second Sunday following adjournment, when congress would be out of town.

* * *

On this trip to the studio, there were no enemies. The Speaker, Senator Meyer, Reverend Rowan and others were all home with their constituents.

Congress had adjourned until January 2 for a lengthy holiday vacation. Labor Day, Thanksgiving, Christmas, New Years and the need to relate with constituents were used as an excuse for dumping a litany of problems on the President. Because of the budget vetoes, a vengeful Congress left Washington without extending legitimate appropriations to cover vital agencies. They expected such tactics would force President Ericson to call them back into session to avoid chaos. Speaker Metzer knew blackmail had worked before on stubborn presidents.

Tony Pellini, Senator May Trudeau, Air Force Chief Pete Hardin--May's constant escort since her husband's death, walked with Jodi into the studio.

So why was she so nervous?

Jodi replaced the mirror in her purse after checking her hair and lipstick for the umpteenth time. She knew the answer.

Her next step had never been risked by any other President in history, with the possible exception of Abraham Lincoln. And that was during a war.

After she breached the barriers there could be no turning back. Whether at Waterloo, Arbela, Trafalgar or Gettysburg, there came a time when the progression of events removed choice.

"We are all set," said Marvin Pringle, stepping forward and taking President Jodi Ericson by the arm to enter the studio prepared to Jodi's specifications.

Jodi had expressed preference to stand before a podium. Although past presidents chose to sit behind the desk in the Oval office, standing gave her a commanding position and the studio furnished a live responsive audience. The large attendance for Pringle's weekly national show provided feedback and a sense of reality.

The podium had only one microphone. Jodi waited for the red light to come on.

"Greetings, fellow Americans."

Jodi suddenly felt the tension drop away as she glanced down at the script and immersed herself in the solid subjects on which she had spent so much time.

"I come before you with the heavy heart that each of us has increasingly shared over recent events. The death of two noble men--President Roland and my husband--and the tragedy of a congress which has abandoned its responsibilities to the American people.

"President Roland laid out a course of returning power to the people and lifting the crushing burdens reckless spending programs saddled on families. Imperial Congresses in the last two decades have created more bureaucracy, increased spending, piled on more debt, then added new federal and state taxes which the average American can no longer support. Today taxes take over fifty percent of all the average American earns. The Czars took thirty percent from their serfs, which we thought unconscionable. Pharaohs in Egypt took twenty percent of slave production.

"To insure that all citizens escape from the load of a rapacious bureaucracy imposed by a greedy and self-serving congress, I am placing before you and your state legislatures an amendment to the constitution.

"This amendment consists of a dozen articles. With the cooperation of your governors and legislators I will introduce this amendment to each state. Later, when ratified by three-fourths of the state legislatures, this amendment will become part of the Constitution of the United States."

Jodi took a deep breath and sipped on a glass of water. She glanced at Senator Trudeau's party, watching intently. They smiled encouragement. Now came the dicey part.

"I've put 12 articles under one amendment so congress can't eat away its guts with revisions like it's done in the past. Past amendments came on like wild tigers and ended up toothless domesticated cats.

"Our forefathers arranged this end run to circumvent any obstreperous congress or unresponsive president. Although they provided an impeachment process to control the president, they provided no such arrangement to control congress.

"Just as President Roosevelt did in May of 1941, six months before Pearl Harbor, I'm proclaiming a National Emergency--effective immediately. There's no need for undue concern that I'm embarking on a road to conflict. This is purely an internal remedy to throttle congress with restraints so they'll quit taking money out of your pockets. We want a republic which functions again.

"I plan to tour the country, visiting as many cities and states as humanly possible, to acquaint you with the intent, purpose and justification of each of the twelve articles."

Jodi hesitated, reading from her notes carefully.

"The constitution says the President 'may convene and adjourn Congress...to such time as he shall think proper.' I think the proper time to convene congress again will be after passage of my proposed amendment--not before.

"Congress adjourned of its own volition until January unless recalled by the President of the United States," Jodi said emphatically. "I shall neither make nor permit such recall.

"The curse of congress is they forgot the tenth commandment--*Thou Shall Not Covet.* They covet the wealth of the nation's citizens to dispense as they see fit, never once thinking those capable of earning money are wise enough to spend it.

"Too many voters have forgotten the first thing they owe their children is the ability to survive free: either from atomic bombs or economic bondage.

"Slavery takes many forms. Physical bondage and servitude are obvious. Political submission and domination less so. Congress created laws and regulations which quietly slipped on bureaucratic handcuffs. Cults practice such domination regularly. Ideologies subtly imprison the minds of people. Taxation deadens resolve and hope into a lassitude of listless weariness which settles over a people until they can tolerate the yoke no more. Desperate people turn to bloody revolutions which produce dictators, too late for citizens who should have acted earlier."

Jodi thought for a moment and then turned to Marvin Pringle, who sat quietly measuring the reactions of the audience. If their attention wandered he was expected to jump in. They were absorbed. He signaled Jodi to keep going.

"Parents struggle to provide child care, health care, better schools and ultimately a college education to their children. Then after a quarter-century of sacrifice they blindly throw these same loved ones into a pit of tar: into a black hole of their own making...a black hole of taxation and financial slavery...a black hole of ruinous debt created by congress from which their offspring can never escape."

Jodi glanced at the large clock on the wall. Time for the presentation of her twelve articles. She nodded to Pringle who gestured to the stage crew. They immediately brought in a large TV screen and set up a projection camera. Soon a white background with twelve articles displayed in bright red flashed before the audience.

"With the help of Mr. Pringle I now present Constitutional Amendment XXVIII," Jodi said. "Mr. Pringle has a long pointer which will follow each article as he reads it aloud. After each reading I will offer a quick justification for that article."

Marvin Pringle read the first article, speaking slowly.

"1. Only one project at a time will be included in a bill for the President's signature or veto. Such bill shall not exceed 10 pages. When not possible, the President, at his discretion, may invoke the line item veto before signing a larger bill into law. No bill over 10 pages shall be submitted to members less than a full working week before vote."

Jodi responded. "Citizens for decades have recognized overload and speed are the most pernicious tools in the tax and spender's arsenal. On legitimate and necessary bills, personal pork items are fraudulently loaded on like barnacles. On their own merits such parasites would never survive the light of day. And since the entire constitution took less than 10 pages of plain English, no law should take more."

Jodi nodded to Pringle who continued.

"2 *.A two-thirds vote of the senate shall be required to reject any Presidential appointment or any treaty negotiated and signed by the President of the United States."*

Said Jodi. "The President is elected by the people from fifty states. Each Senator by only one-fiftieth of that number. Yet senators have resorted to purely personal and vindictive tactics to thwart a President's responsibilities; too often along purely party lines without regard to qualifications. For any President truly out of line, a two-thirds vote would readily reject a bad deal."

Pringle read.

"*3. Total federal government spending shall not exceed twenty percent of the gross national product (GNP); exception--Social Security, administered as an independent self-sufficient organization. Nor shall Congress raise taxes unless by a two-thirds vote of both House and Senate and signed by the President. Exceptions to either spending or taxing totals are permitted only following a declaration of war by Congress or declaration of National Emergency by the President."*

Jodi said. "We've got to get centralized taxing and spending off the taxpayer's back. Congress won't do it. Top down social engineering ended with the Machine Age. It's a dead concept. Machine Age America no longer exists.

"Today an enormous and diverse Information Age is exploding. Personal computers and communications have evolved and swept across countries like the big bang--complex and spontaneous. No elite head-shrinkers can design or control such a new migration of knowledge. As a capitalist country we need to back off and let the productive people work to their full capabilities. Take congress' greedy fingers out of our pockets and bureaucracy's leaden hands off our people."

Pringle continued reading.

"4. Money raised for one account shall not be diverted, borrowed or loaned to any other account, i.e., money raised for special purposes--such as airports, roads, immigration control, or social security--will not be moved into general operations or any other account. Excesses or shortages must be dealt with by separate legislation."

Jodi looked up and smiled at the audience.

"We need to clamp down on the slippery tactics of diversion. Part of the reason congress gets away with murder is that they keep sliding phony chits under the door to each other. Houdini couldn't find where the real money is--when there is any!"

Marvin Pringle raised his eyebrows as he read the next article.

"5. Universal conscription, involving all male youths eighteen through twenty years of age and such female youths as congress might later prescribe, shall take effect July 4 of the year following ratification. Term of service shall be for three years upon graduation from high school or starting from the eighteenth birthday. Only disabling medical exemptions will be permitted plus other personal restrictions, such as felons, as set forth by majority vote of both House and Senate and signed by the President. Furthermore, since all male youths shall be required to serve their country after graduating from high school, all voters must be capable of reading and writing English at that same high school level--the level required of the nation's soldiers."

Jodi took a deep breath, letting it out slowly. This was the article which undoubtedly would raise the most ire with the youth across the nation if it weren't explained and justified adequately.

"We finally get to the hub of the nation's problems. Do we continue with daintily dipping our toes into the waters of superficial social programs or do we immerse the whole body in order to teach it to swim?

"My Youth Work Camps and Universal Conscription, which I will detail in talks across the country, address the roots of our problems. A civilized nation's got to be one in which youths communicate naturally with each other: using the same language, the same currency, receiving equal educational opportunities, wearing the same clothes--thereby learning that clothes alone do not make a man, learning to perform equal chores and cooperating to achieve a common purpose.

"We need to build a strong integrated America. Great leaders have always known that. Abe Lincoln said 'A house divided against itself cannot stand.' Aesop--two thousand years ago--pointed out how one stick can easily be broken, but bundled together they are practically unbreakable.

"We don't build a strong union by penalizing our youth and sending them to prison after the failure of futile social schemes thought up by lawyers, judges, professors and parasites who've never manufactured one article of clothing, built one single house, met one payroll or produced one item of use for the American people.

"Strong nations need healthy, educated and disciplined youths who have learned to work and communicate with each other. Where better places to start a solid foundation than Youth Work Camps and a Universal Military?"

As an afterthought, Jodi added.

"This effort will be financed within Defense."

After a short pause as Jodi sipped on a glass of water Pringle continued.

"*6. Following ratification Congress shall reduce real spending--except as necessary for defense--by ten percent each term to achieve a balanced budget. To do this no new taxes are authorized. Congressional salaries shall be reduced ten percent each term the spending reduction is not met until total federal spending falls to twenty percent of GNP. Defense should stay constant at five percent of the twenty percent.*"

Said Jodi. "This article is self-explanatory. But it introduces a bit of economic caning into a body which has run wild. They've been fed the carrot too long. Now it's time to introduce the stick. I'm sure when our elected officials face real pay cuts they'll suddenly find enough fat in the bureaucracy to provide national defense for another 100 years."

Pringle said.

"7. Terms of office for both elected and appointed officials shall be as follows: (a) Representatives shall be elected for three years. (b) Senators shall be elected for six years, with one-half standing for each election with members of the House. Any interim appointments shall be for the term only of the replaced member. No representative or senator shall serve more than eighteen years. (c) The President's term shall be six years, so he will stand for election with all representatives and one-half the senators. (d) Supreme Court Justices and all Federal Judges shall be appointed for eighteen-year terms. No Justice shall serve beyond the age of eighty. (e) A two-thirds vote of the Senate, signed by the President, may remove any federal judge without cause or prejudice."

Said Jodi. "Much of the reasoning behind this article has been discussed for years. Voter turnout would jump dramatically if half the senators and all the representatives were elected on the same ballot every three years. The President's term would coincide with half the senate's at each election.

"Judges' terms should be reduced to 18 years. That is long enough to provide protection. Our judicial system has lost its way. Our forefathers jumped from the frying pan into the fire. Kings removed judges on whim. In trying to remove judges from the erratic control of monarchs, writers of the constitution swung the pendulum too far in the other direction. This rights that over-reaction."

Pringle changed the pointer to his left hand, then said.

"8. All U.S. Government agencies, organizations and quasi-government organizations will abide by due process, and are enjoined from placing requirements on state and local governments which require mandatory expenditures by states, local governments, jurisdictions or the people without complete funding or their consent.

Burden of proof for alleged violations shall lie with the accusing agency. Miranda-type warnings will be given in all civil as well as criminal contacts by the IRS, the EPA, or any agency of government."

Jodi commented. "The right of seizure, confiscation, or forfeiture of property without due process is no more the right of the Internal Revenue Service or police or drug or custom or environmental agencies than it is of any citizen--no matter how rich, well placed or powerful. Mandates are an unconscionable deceitful means congress uses to load more taxes on the people. Con artists in federal penitentiaries are no more dishonest."

Continued Pringle.

"*9. Gerrymandering shall be prohibited. Congressional election districts shall be constructed for geographical compactness and ease of access. The Supreme Court is to oversee alignments."*

Said Jodi. "Gerrymandering has been hashed and rehashed by the media and the public for decades. It's time we put the issue to rest. Political messiahs have made a career out of hollering that today's human problems are due to other people's unjust treatment. It's a crutch that keeps cripples forever crippled. We need to build all human capital so the borders between peoples' abilities are indistinguishable. Gerrymandered fences don't integrate; they separate."

Pringle read on.

"*10. No federal agency, except those mandated by the constitution, shall live beyond eighteen years without being recreated. A new law passed by Congress and signed by the President shall be required for further funding--after a thorough re-analysis of that agency's value to the nation."*

Said Jodi. "One obstacle to reducing the budget is the innumerable fiefdoms hidden in the enormous federal budget. Like a huge warehouse which is never inventoried, hundreds of hidden bins larded with goodies provide plunder to the unscrupulous and eat away the nation's resources. This article is designed to force such rotten spots out into the sunlight.

Marvin Pringle dropped his pointer to the next item.

"11. Any state legislature, or citizens with a petition bearing five percent of the registered voters who voted in the last Presidential election, may put on the ballot a proposed constitutional amendment. If passed by a majority vote in three-fourths of the states within six years, the proposed amendment shall become law."

Said Jodi. "The referendum process should be more open to the people. Congress too long has thwarted ordinary citizens from participating in needed reforms. This is an effort to bring the people closer to their government. There's no need to fear hasty mob action since adoption of any amendment will require approval by three-fourths of the states."

"Now we come to the final suggested article of amendment," said Pringle, getting up and walking over to the projection unit.

"12. The matter of pay: The federal bureaucracy shall be restructured. Explicit jobs will pay explicit wages, which scales may be reviewed every six years after each Presidential election. Then increases or decreases shall be by majority vote of both houses and signed by the President. Interim increases in pay shall be the result of promotions or transfers to higher positions, not automatic increases. Those above grade GS-14 are considered

political appointees subject to removal without cause or prejudice. Military and federal retirees shall not receive total federal retirement benefits higher than their active service pay in any single agency. Maximum pay in all U.S. Government agencies, organizations and quasi-government organizations shall not exceed triple the average pay of all U.S. workers.

Members of the House and Senate shall not receive one penny over the salaries paid them by the people. Exceptions will be in cases involving their own money, which assets they must have owned before taking oath of office. Nor are members to receive money from individuals or companies which directly or indirectly do business with any government agency--federal, state or local.

Campaign funds will be kept in separate audited accounts."

"Government today faces a pension boondoggle of serious proportions," said Jodi. "If government employees received pensions equal to that of private employees, taxpayers would save $60 billion per year--an incredible amount. Enough to reduce individual taxes by 8 percent. Furthermore, 'spiking'--padding the final three years with overtime, postponed vacations and other goodies--adds billions.

"Such fraud is possible because congress listens to the voting block of bureaucrats more than to taxpaying private citizens."

President Jodi Ericson waited a full minute in the deadly silence which followed before explaining.

"In the past some of these articles have been addressed piece meal. This amendment will supersede any conflict."

When President Ericson finished speaking those in the studio sat stunned. In the audience many congressional staff members squirmed uncomfortably. They could hardly wait to get in touch with their bosses, who they knew would be furious.

Marvin Pringle stepped into the tense atmosphere. He shook his head with a worried expression. He knew Jodi's proposal was a call for all hell to break loose.

"A large order, Madame President," he said, rubbing his hands thoughtfully. "A very large order."

"A very large problem faces the people," said Jodi.

A smile gathered at the corners of Pringle's mouth. One thing for certain, ratings would zoom into the stratosphere if he could get President Jodi Ericson back on the program after her declaration of a National Emergency hit members of congress.

"Perhaps we can explore your constitutional amendment ideas more in depth on a later program?"

"Of course," said Jodi. "People are entitled to the reasons behind each article. I've got a lot of work to do. Starkly presented, these articles are quite formidable and probably frightening to some. Yet when we get down to individual analysis, I'm sure the people will see each article as necessary to a functioning democracy."

Marvin Pringle received a signal from the booth.

"We look forward to further discussion," he said with a broad smile. "Right now we're out of time."

32. WHITE HOUSE PRESS CORPS

Against the advice of Senator May Trudeau and Tony Pellini, Jodi agreed to the demands of the White House Press Corps for a press conference.

"You're breaking your promise to conduct all media sessions outside Washington," protested Tony.

"Promises made only to myself," said Jodi. "Clamor of the past four days is incredibly strident and growing. Best nip it in the bud. Congressmen descending on the town are staging all sorts of impromptu parleys...spreading lies, asserting my National Emergency is the first step towards dictatorship. I learned on the farm to stamp out poison weeds before they take root and spread."

"Newsmen follow dissident congressmen and hang on every word," said Senator Trudeau, sipping on a glass of red wine. The trio was having lunch in the White House. She raised her glass in a gesture of defiance. "With congress no longer in session and the print chasing the baying hounds, it's nice to be left alone."

"Euphoria won't last long," grumbled Tony. "Demagogues gather artillery to blast the President on impeachment."

"I'm still in charge of the Armed Forces," Jodi said, grimly wondering if she'd have the guts to deploy them if needed.

* * *

In the White House Press room President Jodi Ericson surveyed the rows of media personalities champing at the bit. This time they really had something newsworthy--an historic confrontation between a President and Congress. A beautiful female sacrifice for tabloid covers? Surely a neophyte woman President would be tasty meat for tough experienced old media pros. The press anticipated a morsel about to be cut into spicy tidbits.

TV cameras with their zoom lens moved quietly into place. They were positioned to capture any intimate moments of weakness.

Detecting the restlessness of the Press Corps, Jodi toyed with her notes on the podium, building the tension.

"Good afternoon, ladies and gentlemen," the President said amicably. "Since I'm standing before the most knowledgeable group in Washington, I suspect there's no need for a background briefing on my proposed constitutional amendment. So I'll start right off with the questioning."

The assembly laughed heartily. The tension broke. Jodi pointed to *The Washington Post* for the first question.

"Your high-handed National Emergency declaration was pulled out of thin air," Ellen Corey read from her yellow pad. "What gives you the authority to flaunt the constitution? Are you embarking on an era of scorning the laws of Congress?"

Ellen Corey primly sat down with a smug look around the room. Her daring would make her the hero of the cocktail hour with her consorts.

President Jodi Ericson smiled. She had expected tough hostile questions from the media.

"The Constitution empowers the President to 'convene both houses, or either of them, and with respect to the time of adjournment, he may adjourn them to such time as he shall think proper.' Of course," Jodi added, "that is in case of disagreement.

Since the congress can't agree on anything, I presumed they disagreed on this. And if you've done your homework, you know the Declaration of a National Emergency is not without precedent. Roosevelt did it; Abraham Lincoln acted without bothering to declare one. So I've decided there's no other way out of the gridlock for the American people."

"Speaker Metzer warns you're another Czarist Catherine the Great," said *The New York Times*. "He thinks you got in by a fluke."

"An elected fluke," corrected Jodi. "The Speaker's unhappy his protégé Reverend Rowan wasn't selected Vice President at the convention."

"Just what do you expect to accomplish without the help of congress?" asked the *News*.

"What any race horse expects without a 250-pound jockey on its back."

"You think congress a load?

"Too big even for a tractor," the President said with a smile. After a moment of silence she continued.

"My Constitutional Amendment carries twelve articles which I set forth in my television address. Your papers carried and debated the issues all this past week. By this time you undoubtedly are quite familiar with them."

"Will you keep congress out of their offices once their vacation adjournment is over?" queried the *Associated Press*.

"I've alerted the National Guard. I haven't made that decision yet. The National Emergency takes effect whether congress is in session or not."

"We attended a bitterly hostile gathering yesterday at the National Press Club," the *News* stood up to announce. "Jesse Rowan, Chairman of the House Judiciary Committee, is working with the Speaker to draw up articles of impeachment against you. They appear deadly serious. Are you worried?"

"I quit worrying when President Roland and my husband were assassinated. Now I feel like Farragut. 'Damn the torpedoes, full speed ahead.' I lost something more precious than my own life. What more have I to lose? Yet Americans have everything to lose. Submitting like beasts of burden to taxes piled higher and higher will eventually break their backs."

"The moment congress adjourned you took a series of actions contrary to the known wishes of congress," spoke out the *Tribune.* "Can't you possibly cooperate?"

"That idea is a media myth. Presidents are elected to serve the people, not cooperate with Congress," said Jodi. "History has proven the two incompatible. Washington would have failed if he'd turned tail and fled with the Continental Congress; Lincoln ignored congress or he'd lost the war; Roosevelt gave 52 destroyers to save England from Hitler despite congress, and Reagan brought down the 'evil empire' by fighting congress every step of the way. Congress has not been a lightening rod of leadership."

"Nonetheless," said the *Tribune,* "our forefathers established congress with sole legislative responsibility. Why can't you compromise?"

"Our forefathers established three independent branches of government," said Jodi. "They were not meant to always compromise. Otherwise they'd have set up a monarchy."

"Then how do you expect to get laws passed?"

President Ericson smiled at the *Tribune's* dogged questioning.

"Like the physician," she said. "First do no harm. Don't hastily operate or prematurely pass flawed laws. Each does considerable harm to innocent citizens. We've too many bad EPA, EEOC and other punitive laws on the books now. They've permitted a thousand percent rise in malpractice and litigation costs. An explosion of benevolent sounding programs has resulted in a hangman's noose for innocent victims. We failed to 'first do no harm.'"

"You think your amendment will cure all the ills?" *The New York Times* questioned sarcastically.

"Major ones," said President Ericson, fixing her eyes on the *Times* until he turned his face away. "Sometimes it takes a painful needle to cure the patient."

“You mean,” asked an alert *News*, “you intend to extract more painful cuts from congress?”

“If possible, I’d abolish congress altogether and replace it with a new body,” said Jodi. “It’s over 200 years since our forefathers dumped the Continental Congress for incompetence. Now it’s time to dump this one for the same reason.”

“Why?”

“No supervision...no self-discipline. It spends other people’s money without restraint. Our forefathers envisaged elected representatives to serve them. Instead Congress became inbred royalty...voting themselves and their lackeys higher salaries than 90 percent of the persons they were elected to serve.”

The NBC TV anchor man in the front row raised his hand. President Ericson nodded.

“Actions you’ve taken in congress’ absence,” he asked respectfully, “may I go over some of them with you?”

“By all means. I want the people to understand.”

“The $5 billion allocated for job retraining, mostly youths being separated from the military, you’ve abolished completely?”

“Right. Job retraining to what? Unemployment lines?”

“Under President Clinton, Labor Secretary Reich...”

Jodi interrupted contemptuously.

“Another Rhodes lawyer who never produced a broom, ran a business, or met a payroll. His idea of reform would have been hooted down in *One Flew Over The Cuckoo Nest.* During Secretary Reich’s tenure there were 125 different employment and training programs costing $16 billion per year. More than 75 different anti-poverty programs cost taxpayers $210 billion a year: more than twice as much as giving the poor outright cash to lift them above the poverty level.”

“You’ve a better solution?”

Jodi brushed hair from the nap of her neck.

“As TV anchor you surely noticed the pictures of young men and women notified of forced retirements in our Armed Forces? Tears streamed down their cheeks.”

“We all felt their pain,” said the anchor.

"A pain absolutely unnecessary," said Jodi. "I've ordered all such honorably dismissed youth reinstated in the military. Also I've ordered cancellation of 200 superfluous pork retraining programs. We'll use job retraining money to keep young men at their old salaries and ratings...retrain them while still in the military. A far less costly deal than spending $5 billion on one program for stupidly teaching them to become welders, unemployed college students, or social workers."

"You don't believe in job retraining?"

"Job retraining by the federal government is an immoral hoax perpetrated by misguided crusaders who never held a job in the real world," said Jodi. "Another wand waved before the public's eyes. Money which sets up more government bureaucracy, trickles through layers of lawyers, administrators, social workers, unqualified instructors and finally ends training for tasks which by then have become obsolete in the market place. When the jobs are there, Americans always find ways of filling them. Let's create the jobs. Who industry needs, industry will train. When that happens, we will help. Thousands of top executives started out at the bottom of the ladder."

The Star had a question. "Crime is a major concern across the nation today. How would your amendment help?"

"That problem will be addressed under the National Emergency edict," said the President. "I'm establishing a Youth Work Corps, a YWC for all youths 14 to 18 who are not in school, including those who are expelled from school. Run by the military, youths will work, study, and responsibly be reintroduced into society."

"Another Roosevelt Civilian Conservation Corps?" asked *The Star*.

"In structure yes," said Jodi. "But far different internally."

"How different?"

"Schools will not be forced to keep trouble makers. Teachers should not have to cope with bullies, disruptive or mean kids. Brats should be kicked out. Let teachers get back to teaching the 95 percent who want to learn. Yet those outside the school system need someplace to go."

"A boot camp?"

Jodi smiled at her tenacious questioner.

"There's one thing potential young delinquents hate worse than the threat of incarceration." She paused, enjoying baiting the stubborn reporter, who, she admitted to herself, asked all the right questions.

"And what do they hate most?" His annoyance began to show.

"Work. Plain labor. They've learned on the streets they don't need it. They turn to easy crime which takes no longer than an hour per day. To avoid work they're willing to rob, bully and intimidate others. Quick crime gives dropout kids hours and hours of leisure time to impregnate young girls, rap with buddies, tour malls or sneak into alleyways to do drugs. A long hard day's work and study in a military atmosphere does wonders for a good night's sleep."

"After 18? When they go back on the streets?

"Evidently you overlooked my article of Universal Conscription. They next go into the military for three years."

"Congress will yell to high heaven about restricting individual freedoms," persisted *The Star.*

"Congress yells about everything," smiled Jodi. "I bought ear plugs. Never before in history has a body of 535 adults failed to grasp what any 10-year old knows. You can't buy candy when some bully on the street takes money away from you."

"Where are you going to get the money to pay for all these work camps and universal conscription?" *The Star* cried out.

"From those granaries full of rats and their nests," said Jodi. "Fumigate and remove the well-stocked perk sanctuaries from the federal budget and the resultant outpouring of money will amaze you."

"Such as?"

"The Superfund, a poisonous bit of litigation costing over $30 billion, 80 percent of which goes to lawyers and related costs. Estimates over the next decade reach $1.5 trillion."

"Superfund is part of the Environmental Protection Administration," cried *The Post*. "Would you abolish that too?"

"No. But I intend to sharpen its mandate. In connivance with some judges who are a disgrace to their profession, the EPA has sent innocent men to prison, fined them for disagreeing with outrageous seizures of property without due cause and succeeded in usurping powers we refused King George."

"You think the EPA worse than King George?"

"In some cases even more so," said Jodi. "Unelected and uncontrolled bureaucrats have seized power and the purse since the days of the Roman Empire. Emperor Tiberius' had to execute his Chief Minister Lucius Sejanus after trusting him with too much power. That bureaucrat got to thinking he was bigger than his boss. Well, the EPA now thinks it's bigger than its boss--the American people."

"They're arrogant know-it-alls," shouted the Des Moines Register.

Said Jodi. "Thousands of counterproductive regulations need to be banished. Bureaucrats who usurp legislative authority by creating regulations out of thin air should be fired. This I intend to do as fast as they are brought to my attention."

Ellen Corey of the *Washington Post* stood up again. Flushed from her first brave action, she demanded.

"Members aren't going to sit still while you gut their districts of needed projects. Aren't you needlessly starting a fight you can't win?"

"The colonies started a fight they weren't supposed to win," snapped Jodi. "Incumbents spend other people's money on their own districts. That means raising taxes on the rest of the nation to take care of number one."

"The American way," smiled Ellen Corey.

"Perhaps. But destroying other people eventually strikes home. Soon they're left standing on an island of pork surrounded by quicksand. Not an attractive place to live."

Ellen Corey wasn't to be put off.

"There's no way you can stop hog-piling," she said smugly, referring to the practice of attaching junk spending to popular or necessary legislation. "Committee chairmen pass down legislation just twenty minutes before vote. There's absolutely no chance of members reading it."

"Reason to pass my amendment," said Jodi. "Legislation creating Social Security was only 36 pages long. The original health care reform bill under Clinton was 1,342 pages long. Congress has become a stockade full of untamed animals incapable of governing."

"Seems to me there's little an executive can do to control the independent actions of congress," said the *Washington Post* reporter before sitting down.

"You may be right," conceded Jodi. "But too many presidents let agencies forget their boss. Bureaucrats appease liberal spenders in congress who handle their budget, often ignoring the President. I intend to correct that line of authority. Americans spend over $500 billion annually--$5,000 per household--complying with federal regulations. It's time to apply a halter."

The Times waved his hand without rising from his seat.

"Yes," acknowledged Jodi, pointing.

"You've not mentioned college," said the Times. "A great deal of federal money goes to universities and student loans. How do you intend to curtail those costs, which rise like NASA rockets?"

"I've got my Secretary of Education drawing up a new set of rules."

A murmur of concern filtered across the room. The media had been on a highly visible campaign prodding congress to increase funding for education.

"I'm withdrawing federal funding from all colleges and universities when faculty does not average 20 hours per week in the classroom."

Jodi held up her hand to quell the tide of vocal protest.

"Twenty hours per week does not include a three-month summer vacation. I doubt faculties will be overworked.

"I do not think it unreasonable for professors to meet face to face with the people who constitute the reason for their being--students."

"You'll create a war on campus which will make the sit-ins and demonstrations of the 60's look like idyllic picnics," shouted NBC news.

"Productivity is a dirty word when it comes to education," declared Jodi, ignoring the rumbling in the back corner of the room. "Many professors lecture by rote from the same notes year after year. After all, Napoleon hasn't fought any new battles, Newton discovered any new laws or our own revolutionary and civil wars produced any new leaders. During the year, professors now teach students little better than one day a week. That labor so fatigues them they take off three months in the summer...And they demand more pay? A panhandler has as much right to more money.

"Urban blight and housing problems are worse now than before the creation of the Department of Housing and Urban Development. Education results worse than before the creation of the Department of Education. Welfare dependency greater now than before a decade ago."

Jodi continued. "We've succumbed to a national hypocrisy. We cry for better education but tax parents on the money it costs to send kids to school. Then we hand educational institutions millions from that same money to grant scholarships and dole out funds to attract students."

Jodie looked over the White House Press Corps. They were becoming restless. She had given them enough gist for a bundle of headlines.

"I've time for one more question," she said, arranging her notes on the podium preparatory to leaving.

The Washington Times, quietly attentive until now, raised his hand.

"Your 18-year rule for reevaluation of all agencies imposes a tremendous amount of work...for what? It doesn't make sense."

"Hundreds of agencies are hidden in the cracks of our mammoth federal bureaucracy," explained Jodi. "There are sub-agencies to sub-agencies. Agencies grow and extend tentacles to states, counties and cities across America. They fatten like hogs. The bigger the bloat, the higher the promotions. They buy or rent additional office space, issue more regulations and expand into any crevice where they find no resistance. The biggest single payroll in many cities is government."

President Ericson spoke with lips trembling in anger.

"A budget package no congressman can read hides vermin. As I've stressed so often, a huge grain elevator not regularly cleaned and fumigated becomes a permanent home for rats. Time taxpayers stopped feeding rats."

"How about impeachment?" shouted a voice in the back. "Do you think you can survive impeachment?"

President Ericson stepped up to the microphone.

"To shoot someone in the back, you first need a gun. I intend to take that away from congress?"

"How?"

"The way any commander in full charge of his troops does until relieved. I'm still President. Still Commander in Chief."

"Actions of a damned dictator," shouted a voice at Jodi's departing figure.

After a moment's hesitation, Jodi returned once again to the podium.

"Something I forgot," she said pleasantly. "As you leave, please turn in your White House passes. Assistants are stationed at the exit to accept your cards. Future admissions may be obtained by phone. Thank you."

33. ARTICLES OF IMPEACHMENT

Speaker Metzer, Senator Meyer, Congressmen Hockaday and Rowan were in a blue funk. They were indecisive about which three articles of impeachment should be broadcast. They needed to act before the President's National Emergency rallied the people behind her. They fretted over what happened to Parliament when Yeltsin grabbed the initiative.

Speaker Metzer ran his fingers wrathfully through his head of thick white hair.

"She's got to be throttled...and quick. Swing media focus from her constitutional amendment proposals to her usurpation of power."

"People place congressmen lower on the totem pole than used car salesmen," said Huey Hockaday, representative from West Virginia.

"Not half so low as a despot," retorted Jesse Rowan. "Make her into a villainous Catherine the Great or hateful rich brat. That will deflect anything of substance she's got to say about us."

"A trap she's walking straight into without our help," said the Speaker.

"Let's get down to specifics," insisted Senator Meyer. "I've researched the two other instances of impeachment: those of Presidents Andrew Johnson and Richard Nixon, although Nixon resigned before the House vote."

"You're from the Senate," protested the Chairman of the House Judiciary Committee. He didn't want any senator invading his turf. "It's the House's job to draw up articles of impeachment."

"I know that, Rowan," Senator Meyer responded in scathing tones. He had little use for the rabble-rousing preacher who grabbed any microphone, marched in any parade or appeared on any TV show where he could get a bit of self-promotion. As head of the House Judiciary Committee, Rowan would handle the show. There was no alternative to working with him.

"Eventually the President must be tried before the Senate," Mordecai continued in a softer voice. "So we want to be sure the charges will stand one hell of a lot of scrutiny."

"Right," said the Speaker. "The articles got to win the approval of both houses."

The Speaker shared Mordecai's opinion of Jesse Rowan but controlled the House by facing realities, not indulging in perfect solutions. The black caucus, which Rowan led, had grown into a force of fifty members. Mostly they voted in a block.

"President Andrew Johnson fired Edwin M. Stanton, Secretary of War, without notifying the Senate, supposedly and vaguely repudiating the Tenure of Office Act," said Mordecai.

"An arrogant ass," said Congressman Hockaday. "Stanton should have been shot instead of fired."

"Agreed," said Mordecai with a grin. "Congress later repealed the Tenure of Office Act. But the impeachment efforts on both Johnson and Nixon exposed one thing."

"What?" queried Congressman Hockaday.

"Legal charges aren't critical. Making the victim into a tyrant, a cheat, a liar or just plain son-of-a-bitch is far more important."

"Which can be done with any abrasive personality if handled right," said the Speaker. "Look what they did to Leona

Helmsley by calling her Queen of the Mean. She was convicted before she went to trial."

"Enthusiastic backing by the media is a certainty when the jackals smell blood," said the Reverend Rowan. He had led the press down many flights of aggrieved wrongdoing, capturing prime time.

"Things haven't changed since 1868," said Hockaday morosely. "Radicals were determined to get rid of Andrew Johnson and implement their own harsh reconstruction plans. The vindictive Senator Charles Summer and the malevolent Representative Thaddeus Stevens merely wanted President Johnson out of the way. The press joined wholeheartedly in the assault."

The Speaker glanced at Hockaday questioningly.

"You with us or against us?"

"Pointing a few salient truisms about how easy it is to incite the press. Facts won't win an impeachment."

"What will?" the Speaker snapped.

"Emotions. Articles of impeachment, to succeed, must be constructed to appeal to the emotions."

"Right on," cried Reverend Rowan. "My congregation is bored unless I bring tears to their eyes. Then they hoot and holler and stamp their feet with such fervor they don't feel a thing when I pass the collection plate."

Senator Meyer took a folded sheet of paper from inside his coat pocket. Studying his notes for a moment he read.

"My suggested," he bowed with exaggerated diffidence to Reverend Rowan, "three articles of impeachment are:

"1. Obstruction of justice by impounding money and unlawful firing of persons in government agencies.

"2. Abuse of power by unlawful declaration of National Emergency.

"3. Contempt of Congress, a body created by the Constitution."

"Vague," said Hockaday.

"With deliberate intent," retorted the Senator. "As you pointed out, this will turn into a media battle of emotions."

Huey Hockaday's look at the others was ambivalent. Feeling little comfort at what they were proposing, he raised a nervous hand to his forehead.

"Nixon faced three articles charging him with obstruction of justice, abuse of power and contempt of congress. Looks to me like you merely copied them?"

"Go with what works, eh," smiled Mordecai. "Those articles forced Nixon to resign. Like you said, substance isn't important. We've got to get that fancy young lady to step into our pit of emotional quicksand."

"Excellent," exclaimed the Speaker. "I'll get hold of "Jam" Johnson and release our proposed articles of impeachment to the press." He chuckled, his fat jowls shaking with amusement. "Get the bait out in time for the nightly TV news."

* * *

While her adversaries were drawing up articles of impeachment, President Ericson was speaking to a gathering of veterans at their yearly convention in Washington. Arrangements had been made by General Hardin to address a packed luncheon in the cavernous ballroom of the Sheraton Park Hotel.

As usual she was taking questions from the audience.

President of the Springfield, Missouri, lodge, an ex-Marine Colonel, spoke up for an incensed group. He was a tall, slender, stiffly erect man with the weather-beaten face of a farmer.

"We and our forefathers fought like hell for freedom," he declared in angry tones. "Now every time we turn around congress tells us what we can and cannot do. Bureaucrats out-do Salem witch hunters when it comes to looking for microscopic things to remedy. Then they levy a tax to create an agency to fix something that doesn't need fixing. We get more orders and regulations from the government as farmers than we ever got fighting the enemy in World War II, Viet Nam, or Desert Storm."

He stopped and took a deep breath, gazing belligerently at the podium.

"How can we ever get our freedoms back?"

Jodi waited for the deafening applause to die down before replying. She spoke softly at first, her voice rising as she warmed to her subject.

"I see many among you wearing the Purple Heart. You had the courage to face enemy guns so you undoubtedly have the courage to demand back those freedoms."

"Yeah. But how?" cried a man with crutches on the aisle.

"There's only one answer to your dilemma--pick up your guns again. I can help you only so much. Your ammunition is your vote. I find it odd that everywhere I go I hear congress berated. But you're still reelecting too many of the same congressman. Like Stalag 17 the enemy within wears your uniform. You've been blind. Wake up and kick the phony ones out."

"Easier said than done," cried one.

"Vote on their results, not their speeches. You won battles by defeating the enemy, not listening to him. Freedom is not handed to us. We've got to keep earning it. Quit being conned. Accept neither excuses nor promises. Let your local congressman unequivocally understand. You get back your freedoms or they get fired."

"Pretty drastic stuff," hollered a man from the rear.

President Ericson nodded agreement, then added.

"No war's won without a fight. The Republican capture of both houses with Newt Gingrich as Speaker shows it can be done. But they let it slip away again--back into the hands of bureaucrats. Congress can't seem to change its stripes, no matter how often it changes handlers."

"Then let's fight. We've had it up to here," the man cried, drawing a finger across his throat.

Jodi held up the palm of her hand.

"Fight with the proper weapons--the vote," she said. "If you've not the stomach to fight, you can never win. Without the stomach to toss out spineless local representatives, we can't expect democracy to succeed."

34. COUNTER-ATTACK

President Jodi Ericson's percussion cap teetered close to explosion in the next few days.

The press yapped at her heels like hyenas sensing blood from a kill. Each TV anchor sought to outdo the others in a numbers game of votes needed for impeachment. Meanwhile the President struggled to get her message across on the need for a constitutional amendment to curb congress. Jodi often thought back to Yeltsin's confrontation with Parliament. Panicking too soon could be as disastrous as acting too late. Yeltsin let Parliament dig its own grave before acting. Timing was everything.

Tony Pellini offered his support. He read from a copy of the constitution.

"The President shall be Commander-in-Chief of the Army and Navy of the United States, and of the militia of the several states..." He looked up with a wry grin. "You're still the boss."

"This isn't Russia." Jodi wiped her brow with her handkerchief. "Nor was our constitution created by a bunch of hardline communists."

Jodi smiled weakly at the other members of the luncheon seated around a table in the White House dining room. Senator Trudeau, ever the optimist, exuded confidence.

"You're sitting on a powerful military. Only they need a leader. Congress can't command a canoe. Get out into the country and pound your amendment."

Governor Aker, who had flown in from Nevada after attending a Governors' Conference, summarized the mood of the Governors.

"You're the underdog. When the hell haven't you been? You handled my campaign when rabble-rousers invaded Nevada and put me at the bottom of a tar pit on gambling. I don't remember seeing daylight until 24 hours before the election. Yet we won by a landslide. Morning, noon and night you slugged it out in all corners of the state. You didn't win my election sitting in the state capital. Now get out and win your own the same way--in the obscure corners of the nation."

Jodi agreed. "You're right, Governor. I think it's time we stopped thinking about the big picture and stood in the shoes of the man who brings home the weekly pay check."

She turned to her right arm since her husband's death.

"Tell me, Tony. What does the man in the street want?"

Tony downed his drink and pursed his lips. He thought back to the time he was a struggling boxer living from one pay check to the next.

"To keep what money he earns. Getting his brains knocked out either in the ring or the business world is tough enough," said Tony. He sat quietly for a long time running a forefinger around the rim of his glass.

"My first big pay check came after I was over my head in debt. After college I spent two years getting my brains and body hammered into a mass of pain. Finally I won a big fight. I was ecstatic. I could pay off all my debts. But the IRS stepped in and took half my earnings. To compound the horror they reached back and said I owed them for past years plus interest. I ended up with nothing. Agents said they were just administering the law."

"How awful," said Senator Trudeau.

"I found I wasn't the only sucker clobbered by government bureaucrats. They also seized every dime from my handlers, poor ordinary wage earners."

Jodi reached over and covered Tony's hand.

"Earning wages or owning property in this country is no longer a safe haven. It's a highly visible place for demagogues and congressmen to go for more money."

"Where one can't escape scrutiny," added Senator Trudeau. "Everything's a matter of public record, allowing congress to pick pockets dry."

"One more reason for fighting invasion of privacy," exclaimed Tony.

"Better get perking, Madame President," said Governor Aker. "Speaker Metzer and Reverend Rowan set up a phone bank to call all congressmen. You've become a life and death commitment to them. Your proposed amendment threatens their very existence."

"Looks like you're facing an Abe Lincoln confrontation," said Senator Trudeau. "This time a war of words instead of arms."

"Winds of words create firestorms," said Governor Aker. "Witness Lenin, Hitler and hordes of others. Better to repudiate their words before they take root."

President Ericson turned to Senator Trudeau's quiet companion. A strong man with lean tanned features and a hawk-like nose, his brown eyes twinkled with amusement from time to time as he listened, never once offering a suggestion.

General Pete Hardin, Air Force Chief of Staff, for two years had been Senator Trudeau's constant escort. Ever since her husband's death in an airplane crash in the mountains of Venezuela while prospecting oil fields. General Hardin rarely wore a uniform in public. It allowed him to slide unobtrusively about Washington.

"What think you of my program, General? Much of it regards the military."

"I'm no politician, Madame President," he said. "But I do know young men. Your idea of a Youth Work Corps for the young is inspirational. Nothing gives a young man such self-respect.

Watch any parade. Even seventh graders in junior military academies march with enormous pride and dignity."

"The Joint Chief's term expires soon," said the President.

"Next month," said General Hardin, glancing at the President with surprise.

"Senator Trudeau's been touting you for the job," said President Ericson with a broad smile. "Think you can handle it?"

General Hardin blushed, looking over to Senator Trudeau as if he would like to turn her over his knee. This seemed hardly the time or place to discuss such a grave matter. Apparently sensing his embarrassment, President Ericson tried to put him at ease.

"Don't be upset, General. Wherever the President goes, the office goes along. It's the nature of the beast."

"I'd be jumping over a lot of senior officers," said General Hardin. "You'd hear squawks."

"That's for a President to decide. I only want one thing in return."

"Yes?"

"Loyalty. The same kind Genghis Khan demanded from his commanders." President Jodi Ericson smiled. "There's a bit of Mongol still running through my genes."

"I'd never hesitate to give loyalty to my President and Commander-in-Chief. The moment I couldn't, I'd resign." General Hardin glanced at Senator Trudeau who sat in quiet approval.

"Fair enough. I'll announce the appointment in the morning," said Jodi. "I think you'll do a fine job. I'll need strong support in the coming months."

"What about Senate approval?"

"Members of the Senate approve you more than they do me," said a smiling Jodi. "They expect a male military hero might settle me down. With May's strong support I don't think there'll be trouble on that score. Meanwhile, when the Joint Chief's present term expires first of the month, I'll make you Acting Chief."

35. RETURN TO VEGAS

In a quick side trip to Las Vegas--President Ericson preferred Vegas to Camp David--Jodi sat relaxing on the edge of the pool watching Tony perform diving maneuvers off the low board. He was quite good. Young girls flowed around him like topping on an ice cream sundae. He was a fine looking physical specimen; solid arms and shoulders and a tight fanny encased in daring white briefs. A light golden tan highlighted the rippling muscles gleaming with moisture. He moved quickly to Jodi's side and plopped down in a canvas-backed lounge chair.

"Out of condition," he panted. "Ten laps of the pool and a few dives leaves me in worse shape than ten rounds ten years ago."

"You look in pretty good shape to me, Tony," said Jodi. "Those girls eyeing your prowess evidently think so too."

"They're youngsters. Forever young. They've no idea a man eventually passes his prime. Today I'd get knocked out of the ring by an amateur."

"I think they'd like to get in other places than the ring with you," laughed Jodi, tossing Tony a towel.

"Too young. My training days are over."

Jodi got up and walked over to a spot under an umbrella. Tony watched her move across in front of him. A lithe curvaceous body which somehow seemed totally incongruous with being President of the United States. Her blue-green eyes flashed roguishly as she passed, eyeing him from head to foot.

"What's the matter?" Tony asked. "My swimming suit too risqué? I bought it in the hotel shop."

"Risqué? In Vegas?" laughed Jodi. "No. I was surveying the material which seems to have captured the attention of all the young girls."

Tony blushed to the roots of his dark sideburns.

"Children. They fall for images, not reality."

Jodi wasn't so sure. "They seem to indulge in reality at an early age these days."

"Sad reality. Their awakenings are fearful," said Tony.

Jodi fought to ban the memory of her husband from her daily schedule. Not that she had ceased to love him, but dwelling on his memory caused her unbearable heartache. References to him were taboo.

Tony recognized the reason behind Jodi's refusal to discuss anything relating to Guy, his manner of death or his life. Jodi had enough on her hands as President. She had turned the operation of Guy's empire over to Tony, who she both loved and respected for his unshakable loyalty. Fortunately, Guy's demanding personality had enabled him to select worthy lieutenants to run each of his companies. He had always insisted each stand alone on its own profit-or-loss statement and not rely on his holding company for operational or financial help.

Jodi studied Tony thoughtfully for a full minute as he watched a diver on the high board.

"It's important to know someone cares about you--just you," Jodi said, speaking more to herself than Tony. "Not your position, wealth or fame. Only you."

Tony jerked his head around as if he'd been slapped.

"What the hell you talking about? I've watched over you like a Dutch uncle."

"Of course you have," said Jodi, leaning over and kissing him on the cheek. "My thoughts were of my parents. Their struggle that I little appreciated growing up."

"A natural relationship until kids see their parents die. Too soon, too old; too wise, too late. Or something like that. Nothing new."

"I owe you a great deal, Tony. I've often reflected on that fact."

"Hogwash. You've always been your own independent self." Tony grinned. "Sometimes to the point of needing a whack on the behind."

"You came into my life when I was a destitute waif in Washington, fired from a job obtained by political influence, with no prospects of escaping the powerful tentacles of an evil Speaker, and absolutely no hope of obtaining a client."

"Even so, you eyed me as a potential rapist."

"How can I not be beholden?"

Tony's resentment momentarily flared.

"Beholden is a slave word--an intellectual attitude. It evidences no feeling, no love, no emotion. It signifies entrapment."

Tony's vehemence took Jodi by surprise. Evidently she'd touched a raw nerve. Puzzled, she studied her fingernails, glancing sideways at his reddening face. What had she said that so provoked him? Suddenly she knew. Tony was in love with her.

She spoke with quiet gentleness.

"In the life of any human, there can be more than one love. A woman loves her husband. Suppose they have two children. Does she choose between the children? Parents, sisters, brothers each can be loved with all one's heart."

"It's impossible to fight love for the dead."

"Must you forsake one love to capture another," said Jodi softly. "I think not. Did Christ love one disciple more than another."

"I'm speaking of love between man and woman," said Tony stubbornly.

"Second loves may be different, but they need not be less intense." Jodi fought an overwhelming impulse to grab Tony into her arms.

His beautiful body and sincere devotion had been relentlessly creeping into her subconscious mind. Freud was right. While her conscious efforts had been devoted to the demands of her job, her subconscious had been seeking physical intimacy. Her physical needs were mounting.

Jodi reflected with macabre humor that she might be President but she still needed the human touch. Monkeys and wildlife were instinctively smarter than some humans. Always grooming and touching each other. On National Geographic Jodi had been amused at wolf-mates with their mouths in each other's faces. Divorces occur when the touching stops.

"Tony," said Jodi hesitantly, wondering if she weren't overstepping her bounds, "may I ask you a personal question? Frankly you should tell me it's none of my business?"

Tony responded warily. "You may ask. I'm not sure I'll answer."

"Fair enough. But I'm concerned. To my knowledge you've never had a girl friend. Don't you have a sex life of your own?"

"Sex without love isn't particularly satisfying," said Tony. "A bit like indulging in a gourmet dinner without an appetite."

"They're many gorgeous gals in this town."

"None so gorgeous as you, Jodi. You ruined my taste for lesser products. When you married Guy I went into extensive training."

"Meaning?"

"In boxing it means foregoing other pleasures to concentrate on the job."

Jodi studied Tony for a long time. He faced away, again trying to concentrate on the activities around the diving boards. She understood how Queen Elizabeth or Catherine the Great must have felt when they knew no subjects would dare to make a pass. Monarchs had to do the selecting.

"Tony," Jodi said in a soft almost inaudible voice. "Would you like to sleep with me tonight?"

Tony's wide-eyed incredulous expression had the look of a frightened buck facing the barrel of a hunter's gun.

"You playing with me?"

"Not at all. Aren't you in love with me?"

He answered reluctantly with a tortured expression.

"So you know? I tried to hide it for fear it'd mar our relationship."

"You succeeded. I only discovered it a few moments ago because I discovered something in myself."

"What?"

"That love's a two-way street. I lay awake last night wondering what my priorities were? I decided I was in danger of becoming an overburdened pack mule carrying other people's problems with no life of my own."

"I hope you mean it?" said Tony, still shaking his head in wonderment. "I hope you mean it?"

"I do. We'll see if it works out. We'll both be taking a terrible gamble."

"How so?" said Tony, suddenly fearful.

"We can never go back to our old camaraderie."

Tony didn't care. From his point of view the past had been an intolerably strained relationship. Sitting, working, playing next to an untouchable goddess had been living hell.

36. UNIVERSITY OF NEBRASKA

President Ericson stood center stage in the huge coliseum at the University of Nebraska. She had decided to return home to her roots for her first excursion into the nation's heartland. She made a striking presence dressed in cowboy attire. A tight white leather skirt was secured at the waist by a wide rattlesnake belt. A taut yellow blouse strained to keep in check a healthy bosom covered with a loose fitting jacket dangling ropes of braided leather. A beaded Indian headband high on her forehead set off the black hair dropping to her shoulders. Cream-colored cowboy boots completed the ensemble.

Earlier she had asked Tony.

"What do you think of my outfit? Perhaps gaudy, but basically the same as I wore to school most of the time. I want old friends to recognize me?"

"I hardly think they'll have any trouble on that score." He grinned. "As for the outfit, a bit dramatic, but striking."

"I am not a stereotyped cover girl President who happens to be Jodi Ericson," snapped Jodi. "I am Jodi Ericson, an individual character who happens to be President."

Tony scratched his head and smiled.

"I get the drift. Jodi the maverick first. Whatever follows, is secondary?"

Jodi smiled. "After all these years, you've got the picture. Each of us is first an individual. Our later actions define us to society."

"In either case you look beautiful," grinned Tony. "I'm sure the crowd will have no trouble remembering the original."

Jodi chose an open forum in which she would take questions. There had been scads of publicity on her proposed amendment castrating congress, so she believed there was no point in standing and delivering another lecture. University students and military personnel had one behavior down to a science--how to sleep during dull lectures with their eyes open. In contrast, a participating and agitated audience would debate the message long after she was gone.

After introductory remarks by the Chancellor, President Jodi Ericson stepped up to the microphone.

"One thing my undergraduate years at Nebraska taught me," said Jodi, "is that students here recognize when bull shit is the real thing."

A startled gasp went up from the faculty while the assembly hooted with laughter. The remark seemed totally out of sync with the feminine beauty of the speaker. But the people loved it. This speaker obviously wasn't about to hand them more garbage about the higher pursuits of politicians.

"With that in mind," said Jodi, smiling at the relaxed and laughing audience, "I know you don't wish to hear any lies about how well government does its job. So I'll try to answer individual questions. We public officials were elected to serve you individuals, not offer pie-in-the-sky programs created by a congress with its hands in your pockets."

She pointed to a gray-haired man seated on the aisle.

"You have a question, sir?"

Five persons who were selected to break the questioning ice sat in the front row and were tagged with a UN sticker high on their left shoulders. All others were told by the Chancellor they had only to raise their hands once the questioning began.

A man stood up, bracing himself on the back of his chair.

"Madame President. How can you institute reform when the congress is threatening you with impeachment?"

Jodi looked steadily at the speaker for several seconds before answering. It was a brutal question.

"Dear sir," she said to an audience suddenly silenced by the temerity of the remark. "I learned how when I was young girl riding rodeos and state fairs in Nebraska. Once a thunderstorm hit the fairgrounds only minutes before a race. One handler insisted I go back to the barn. 'How can you ride in such a monstrous downpour? You'll be swept off your horse?'

"I told him you win in a rainstorm the same way you win any other time. You mount the horse and run the race. And I intend to run this race."

The applause was deafening.

"Incidentally, in a field of eleven I came in second," she smiled as she hesitated. "It helped that seven dropped out. I think a good many congressmen seeking my impeachment will realize they're headed into one helluva rain storm."

Jodi recognized a weathered man in spotless jeans and a deep hoarse voice. An attendant rushed to hand him a microphone.

"Farmers are constantly pummeled by the big city press for receiving subsidies. We don't receive a dime unless the price drops into the bottom of the barrel. Below the cost of production. This makes it possible for us to risk growing crops which feed the nation. Yet the press and most politicians holler for more food stamps, more welfare, more health care, more housing and a dozen other subsidies which go to people who don't produce one damn ounce of food for their bellies."

He reached into his pocket and produced a worn tattered newspaper clipping.

"The black caucus is from the cities," he said, glancing at an old news story. "McDonnell Douglas rented the cavernous National Air and Space Museum for a reception honoring 14 black freshmen congressmen. Federal Express, Tyson Foods, Chrysler,

Philip Morris, Sony and Pepsi all joined in. The 500 tables at the banquet were sold out at a price of $25,000 each."

The man spun around and faced the assembly with several sour observations.

"In Washington they callously called it 'awareness of minority issues.' Yet not a nickel of $12 million went to a single minority worker. The money went to black congressmen. Here in the midlands we call it for what it is--bribery. But the sad fact is farmers can't compete in the bribery market."

He turned back to face Jodi Ericson.

"That's why, Madame President, our only choice is to rely on a strong executive who will take our part," he said, handing over the mike and sitting down.

After the applause died, the President leaned forward and nodded in affirmation.

"The very reason I'm here. To promote my amendment to the constitution. Congress has lost its way. It no longer serves you people. Fund-raising parties get bigger and bigger. Lobbyists get richer and richer. Congressional pay and perks grow and grow. They now make three times as much--using your money--as you who sent them. Members of that exclusive club want to impeach me."

Jodi straightened and smiled engagingly.

"Frankly, if I were one of them I'd want to impeach me too. For I intend to strip them of their unsupervised conduct. But it's up to you. You have the vote if you've got the purpose. I've offered you purpose in my amendment. Another ten years of congressional leadership may deprive you of that choice."

A short muscular man with heavy eyebrows spoke.

"I run a communications and radar shop in Lincoln. We manufacture and repair sophisticated electronic equipment for satellite, telephone and weather companies. As a small business I've been hit with increased taxes and totally stupid government regulations..."

"Laws passed by congress," volunteered the President when the man seemed at a loss for words.

"Quotas and Equal Employment Opportunities on top of all the other regs are making me throw in the towel," the man cried bitterly. "To stay competitive I, who own the business, had to attend night school to learn computers. Employees in my business are highly specialized. Most know more than I do in their respective fields." He paused to catch his breath as Jodi patiently waited for him to get his thoughts together.

"Take your time," encouraged Jodi. "We want to hear your story."

The man's voice cracked almost into a sob.

"I've absolutely no need for unskilled workers. But government quota regulators decree I must bloat my payroll with worthless workers I can't afford. Stupid...stupid...stupid."

"I understand," said Jodi patiently.

The man continued. "Quotas attempt to put first graders on par with qualified mechanics or computer whiz kids. Like pitting mules and thoroughbreds against each other in the Kentucky Derby. All you'd do is humiliate the mules. Each job demands particular skills, not quotas."

"I've already acted on problems such as yours," said Jodi.

The man looked up hopefully through tear-stained eyes.

"How?"

"Before I left Washington I signed an executive order cutting off the supply of blood and oxygen to numerous regulatory agencies...with more to come."

"Blood and oxygen...?"

"Money," said Jodi. "I've initiated a plan for completely restructuring agencies. News should be hitting the streets next week. Would you care to hear some details?"

A hearty chorus of yeahs gave her the answer.

"Even if it interrupts your question and answer period?"

"We came to listen to you, Madame President," yelled a heavy voice.

"Okay."

Jodi surveyed the packed coliseum. Banners were still in place from a recent school dance. Big bands had been replaced by country music stars. Students loved to congregate for big social events: football, basketball, dances, concerts or political free-for-alls. Except in a few schools run by weak administrators, the students were stimulating...opinionated but respectfully so.

"I've been forced to act by executive order," said Jodi. "Passing any bill to curb congress is not only impossible but would take months. Past experience has shown that new elections mean nothing. Once winners find the pots of honey in Washington, they're like a drug addict finding a fix."

"You've got that right," a voice cried out.

"My first step will eliminate certain agencies altogether. Next merge and reduce others.

"The DEA and FTA, which have been hemorrhaging money with no real results, will be abolished," said Jodi, sweeping her hand across her body to emphasize their ouster. "So will the Departments of Energy and Education, created at a time of panic under President Carter over oil and schools. The energy scare was a bad dream. The Department of Education became hostage to an NEA union of teachers which stuck its fingers into everything except better education for our children.

"Next to go will be the Departments of Transportation, Housing and Urban Development: not exactly critical to the nation's survival. Nor is the General Services Administration--each agency now has it's own purchasing power. Same with the Office of Personnel Management--a paper work factory."

A woman in the front row wearing a badge raised her hand.

"How about crime? We need more money."

"No. I don't think we do," said Jodie. "President Clinton and congress enacted billions for a crime bill which accomplished nothing. Fifty percent of crime today is a mixture of bad laws and bad juveniles. Ninety percent of the prison population is composed of persons who don't commit violent felonies against their fellow man. Yet they're mixed with mean, incorrigible, unrepentant murderers, rapists and vicious criminals."

"You'd not allow criminals to go free?" The woman protested, her face flushing angrily.

"Absolutely not," said Jodi, lifting the palms of her hands in a motion to hear her out. "Real criminals will go to prison. But ninety percent of the vicious crimes are committed by only ten percent who lawyers manage to negotiate through revolving doors. We'll send anyone who harms another to a walled penitentiary. And we'll make them stay there."

Jodi paused, adding. "But we won't mix bad rotten apples with bruised recoverable oranges."

"Rumor says you plan to reactivate military bases?" cried a voice.

"True. Several good ones are in Nebraska," said Jodi.

"When?"

"Immediately. Charles Colson, who spent time in prison for Watergate, once said past crime policies 'have run up an unblemished record of failure.'

"Since 1960 our prison population has quadrupled. We've got the highest per capita rate of incarceration in the world. Yet violent crime has increased 600 percent.

"In a word," said Jodi, "congressional action does less to reduce crime than to increase federal government. Criminals who kill without rhyme or reason have no conscience, no character, no sense of right or wrong and most of all no sense of self-discipline.

"Human beings are born into the world as uncivilized as they were 10,000 years ago. Takes hard work and years of youthful learning to wean us away from automatically becoming barbarians. Youth Work Camps will help that learning process," Jodi said.

"What about the bad juveniles?" A woman questioned insistently.

"I'm activating these military bases for the young ones who most need help."

"I don't understand," insisted the woman, refusing to sit down when admonished by neighbors.

"Juvenile problems start in the homes and in the schools," Jodi patiently explained. The woman had legitimate concern.

"Violent and disruptive pupils make education impossible in our schools. I'm allowing immediate expulsion of such students. Our schools should offer education. They need to return to teaching. Asking a five-foot-five female to maintain discipline in a classroom with a couple of abusive six-foot males is insane."

"You'd put the toughs out on the street?"

"No way. A primary purpose of Work Camps is to handle expelled and truant juveniles. Run by the military, somewhat along the lines of President Roosevelt's Civilian Conservation Corps camps, inmates will study and taste the discipline necessary to good behavior."

"Won't that demand more taxes?" cried an anguished voice.

"Absolutely not. During my Presidency there will be no new taxes. I know you've heard that line before. I don't blame you for being skeptical. But the government rakes in enormous sums of money from you people. It needs to better spend it."

"Amen!" shouted a farmer to loud applause.

"Prisoners cost $25,000 each per year," said Jodi. "Kids in work camps cost less than one-fifth that amount."

Jodi paused, brushing the hair away from the back of her neck where the perspiration made it stick. She grabbed both sides of the podium. Was this the time to also bring in the issue of capital punishment? Didn't she have enough on her agenda? She decided to plunge ahead.

"When dealing with taxes and capital punishment," said Jodi, "there's a sensitive subject which is never mentioned...serial and mean killers with life sentences. We feed, cloth and provide a way of life for fifty years or more for such assassins. At $25,000 per year that's $1,250,000 for the thug who killed a fellow being. With 600 on death row that adds up to $750 million. That much money would furnish a lot of playgrounds and provide hundreds of thousands of books."

Jodi fell silent a moment, disturbed by her thoughts.

"They're like dogs with rabies. I don't hate them," she said quietly.

"You'd execute all those people? A mass murder?" cried a horrified voice.

"People who kill other people are the murderers," said Jodi. "But I'd lower the hype."

"How?" cried the same voice.

"Past debates have missed the point," said Jodi. "Hard liners want executions as a deterrent. Soft hearts think vicious criminals can be rehabilitated."

"And you think not?" hollered a disbeliever.

"I reject either extreme," said Jodi. "Removing rabid dogs doesn't send a message to other dogs. Getting rid of 'Hoof-and-Mouth' diseased herds provides no moral lesson.

"Society can't afford them no matter how pathetic their state."

A deep voice boomed from the back of the auditorium. He had his own microphone. Jodi peered past the sea of faces and saw a huge frame holding the instrument in front of his face.

"Madame President, your whole program is a sham. You're out to distract these fine people from the fact you face impeachment for your own lawless ways."

A tremor shot through Jodi's entire body. She recognized the voice of Ali 'Knuckles' Radman. What the hell was he doing in Nebraska? At once she realized the Speaker had sent him. Ali never moved across the hall without instructions from Speaker Metzer. She glanced with concern at those around him. Ali usually came with a supporting accomplice--a nerd who fed him script. Sure enough. She noted the spectacled head of the Speaker's top researcher, Irving Siegel, who Jodi suspected pulled the wings from butterflies for amusement.

Jodi paled. How could she handle a brutal heckler like Ali who carried his own electrically powered microphone? She couldn't very well direct security to take it away. He was a powerful man and the AA of the Speaker in Congress. The resultant publicity could only be damaging. She decided to face the issue head on.

"Folks, the big man with the microphone is correct. Congress would like nothing better than to impeach me."

A wave of sympathy swept over the audience. TV and the news had been full of President Jodi Ericson's woes.

"For breaking laws right and left," shouted Radman, the sound echoing in the vast coliseum.

Jodi leaned forward into her microphone.

"Before this tirade goes any further I'd like to introduce my heckler. He is Ali "Knuckles" Radman, Administrative Aide to Speaker Metzer. Hardly an impartial critic. How he found his way to Nebraska is a mystery. He must have had help. The last time he came to Nebraska was as a member of a rival football team," said Jodi, drawing on her memory of Radman's resume. "Nebraska whipped him badly. If he doesn't behave, I'm sure they'll do it again."

The rousing acclaim assured Jodi that so far Radman had done no damage.

Radman was as sensitive as a cement post. Also the hate which seethed inside him for the agonies suffered at Jodi's hands made him unstoppable.

"There's no way you can slaughter agencies created by congress and not be impeached," cried Radman. "You belong in prison, dear lady. One day I'll see you there."

"Our Imperial Congress, led by your boss, Speaker Metzer, has usurped many powers," retorted Jodi, her face burning, "but it still hasn't captured the judicial system."

"Impeachment doesn't touch the judicial system. Impeachment and conviction take place in congress," said Radman smugly, lowering the tone of his voice so the reverberations stopped.

Jodi knew Radman spoke the truth. Momentarily flustered, she looked wildly around for help. Radman was about to destroy the very start of her campaign. She was certain Speaker Metzer and his cohorts would be delighted.

Tony Pellini got up from his seat next to the Chancellor. Quickly he approached the podium.

"We're having a bit of problem with the loud speakers to the outside," he said easily into the microphone. "We want everyone to hear. We're switching control boxes. It only takes a minute. Please bear with us."

He covered the microphone with his hand, at the same time whispering in Jodi's ear.

"So he's landed a couple stiff punches. He's an ignorant buffoon following instructions. He'll run out of material by the third round. Stay calm...and stay in there."

Tony lifted his hand from the mike. "They've signaled the problem's fixed," he said with a broad smile. "We can now proceed with the debate."

To avoid Radman, Jodi pointed to a man in the front row wearing the red and white UN sticker on his shoulder.

"You had question, sir?"

"Yes I do Madame President," he said, somewhat startled and jumping to his feet.

"Go on," said Jodi, signaling the attendant with traveling microphone.

"I had a son in the military. He was doing fine. He loved it. He studied avionics, satellite communications, computers, and maintenance on everything from trucks to missiles." He stopped as tears gathered on his cheeks.

"Now he's home without a job," he wept.

"Tell us what happened?"

"The Clinton administration discharged him from the service."

"I know," said Jodi sympathetically. "They were trying to save money by reducing the military."

"Stupidly, they didn't save a damn dime," declared the man, raising his face to the President. "Rhodes Scholar Reich, who didn't learn a thing about American boys in England, came up with a $5 billion program for retraining. Yet that same $5 billion would have kept all the personnel fired from the army. And like my son, they could have kept on learning. The retraining program? Pied Piper Reich led young people to drown in a river of deceit."

Jodi nodded sympathetically at the man's frustration.

"I know. When government moves in to guide youth it's like a rogue elephant directing traffic."

Ali Radman was not to be put off. He revved up his loud speaker.

"Madame President. Don't you understand? You face impeachment. Nothing you say or do has any meaning. Shortly you will be out of office."

"Braying jackasses never yet won a rodeo," retorted Jodi. "The fastest congress ever moved was to flee Washington before the British. The people know how congress lies. Congress voted enormous tax increases. Presidents Reagan and Bush each learned the hard way--congress lies." After pausing, she commented. "President Clinton and congress lied again with another budget buster."

"People want improvements in their lives that taxes bring," shouted Radman.

"Another fabrication fostered by those who spend like there's no tomorrow," said Jodi, tightly gripping the edge of the podium.

How to get rid of Ali Radman was growing into a problem. Continuation of his verbal abuse could derail amendment efforts on her first outing. She suspected the Speaker had instructed Radman to haunt every meeting.

Jodi decided to risk a wild gamble. If she couldn't silence Radman perhaps the huge assembly could. She addressed the crowd with a inviting smile.

"It seems the Speaker's Administrative Assistant wants to indulge in debate. I welcome any discussion of the issues. So if he'll please step up here to the platform we can hear his defense of congress."

The tactic caught Ali by complete surprise. His last words of warning from Speaker Metzer were to heckle but not get drawn into a quicksand of argument with the President.

"She's got a quick, devastating wit which could turn you into a buffoon. Harass, then get the hell out," the Speaker had charged, stabbing his forefinger into Ali's chest. "Guerrilla warfare. That's the ticket. Get into a full blown battle and she'll cut you up and spit you out."

Ali Radman's face reddened at the challenge. He looked about at all the faces turned towards him, and demurred.

"I'm not interested," he said, sitting down.

Jodi announced with a smile.

"Then I'll expect no more interruptions from the Speaker's errand boy and proceed with questions from the audience."

The applause was deafening. Jodi's gamble had paid off.

For another hour President Jodi Ericson fielded questions and ended with meticulously detailing the twelve articles of her amendment. The press had been provided copies earlier, but she was certain the headlines would feature her tilt with Radman. The sabotage tactics of the Speaker had boomeranged badly. Yet she could hardly rest. The setback would only enrage the Speaker to greater efforts of malicious mischief.

That evening, after a few beers, Ali Rahman returned to his hotel room to find a note slipped under his door.

"Dear Ali," it read. "Need another lesson? You're on my home turf now. Your next lesson is scheduled for noon tomorrow if you're not out of town by then. Ever remembering, Jodi."

Ali's dark face turned pale with a mixture of rage and fear. He swung on his roommate.

"We're getting the hell out of this waste land. That woman's a demon."

The nerd took the note from Ali's hand, reading it slowly.

"What lesson's she talking about? Sounds like a bluff?"

"That Mongol bitch doesn't bluff. She's sister to the devil. We're leaving."

Next morning Ali Rahman, fuming but helpless, caught the 8:15 flight to Dulles. In glum silence he spent the entire trip plotting revenge. He had professional friends in New Jersey who were adept at silencing people.

37. RUMBLING IN THE CAPITOL

In Washington things weren't looking well for President Jodi Ericson. While her constitutional amendment was winning support in the field, members of congress were meeting in enclaves all over the city. The White House Press Corps ignored America to concentrate on the juicy scandal breaking within the ten-mile radius of the Capitol Beltway, known as Route 495.

A resolution patterned after the original lodged against President Andrew Johnson read:

RESOLVED: That Jodi Ericson, President of the United States, be impeached of high crimes and misdemeanors in office.

The resolution, drawn by Speaker Albert Metzer and Judicial Committee Chairman Reverend Jesse Rowan, made no effort to detail any charges against the President. They knew, as in the cases of Andrew Johnson and Richard Nixon, that a slew of amendments would come later during debate when congress returned. At that time they planned to insert the three articles which felled President Nixon.

They did, however, inform the media of one charge levied against President Ericson--that she exhibited a "contempt for congress."

Jodi was in San Francisco when she heard.

"At least they got that right," she said to a cheering audience. "If I'd been in Washington I'd have endorsed that charge myself. There aren't enough prisons to hold the people who harbor a 'contempt for congress.'"

The remark evoked thunderous approval and brought the crowd to its feet.

Not forgetting the purpose of her cross-country tour, Jodi nodded to Tony, who stood in the wings of the amphitheater.

Two assistants immediately trotted out a screen and camera. Jodi's 12 articles of amendment were again displayed in red against a background of white. She had been showing them at every stop.

They had been redone by a professional artist so local TV cameras could pick them up with more clarity.

Jodi explained.

"A cadre of your state representatives agreed to submit a constitutional amendment containing these articles," said Jodi. "This has been accomplished in many states I've visited. Three-fourths of the states need to pass this amendment.

"Let me again go over these articles, one by one, so that each individual has a clear understanding of how we want to improve our government. We know our forefathers did not intend to create a legislative body which could reach into our pockets and take hard-earned money for Don Quixote flights of fancy.

"Congress has become so self-deluded they are completely insensitive to the plight of the taxpayer. We mean to change that...by continuing to change the composition of congress if necessary. Over the centuries no body-politic has remained unchanged. Kings, Queens, Emperors, Czars and Dictators all have given way to new ideas. An Imperial Congress should be no exception."

Jodi smiled as she accepted a pointer from Tony and stepped over to the display.

"Please feel free to ask questions at any time," said Jodi. "Our intention is to inform and enlighten. We then hope you will agree with our program for reform."

* * *

When President Ericson finished her review of the 12 articles, a tall gray-haired man in a business suit raised his hand.

"One thing President Clinton inspired and congress increasingly foists on us is the idea of universality," he declared in deep angry tones. "Universal health care, universal quotas, universal day care, universal education. As if anything worth while has to be universal and paid for by universal taxes."

Jodi nodded agreement.

"Because one man needs eye glasses we need not force them on everyone."

"Exactly."

"Politicians often seem devoid of good judgment or common sense," Jodi said. "Mostly it's no more than not knowing what monstrosities they create. They abdicate responsibility and allow the bureaucracy to take over the administration of vague, compromised, muddled laws..."

"Which cause nothing but trouble," the man shouted.

"No question," Jodi said agreeably. "The quest for universal purity brought us Hitler, Tojo, Cromwell and Khomeini. All chasing a common idea of ethnic, religious, or some kind of utopian purity."

"Amen," the questioner hollered as he sat down.

President Ericson expanded the gentleman's theme.

"The Spanish Inquisition sought to weed out the marginally impure and the response was a slide into savagery and bestiality.

"Business and industry also found that chasing universal or 100 percent perfection was ruinous. Manufacturing an airplane, a weapon, an automobile or a kitchen appliance could be improved incrementally as knowledge was gained from usage. Aviation companies fast learned a product could be manufactured to 90 percent of its designed perfection, but trying to achieve that final 10 percent proved fully as costly as the initial 90 percent. And it never worked. Like the Spanish Inquisition, trying to mandate universal behavior signals a stupid and fatal misunderstanding of the human race."

"You've got that right, Madame President," cried a woman in the front row.

Jodi smiled at the woman's fervor.

"Humans evolved from imperfection," Jodi said. "And they will continue to evolve in the same manner. Attempts to cage them into universal care, universal religion or a universal political ideology can only result in catastrophe.

"In a democracy nothing is universal," Jodi declared. "In war the factory worker gets higher pay than the soldier risking his life on the front. Politicians vote themselves higher pay than 90 percent of the people they work for. An alien would stand aghast in disbelief that the servant received more money than his employer.

"The number of efforts in history to achieve some universal utopia is staggering. Millions of lives were sacrificed. The Crusades, ethnic wars, religious wars, freedom wars--in which the people trade one kind of bondage for another--are interminable."

A man on the center aisle stood waving his hand frantically.

"Yes," said Jodi, pointing.

"How about universal health care? President Clinton tried his damnedest."

"A potential atomic disaster waiting to explode if carried much beyond 90 percent," said Jodi.

"Why? The people need it?"

"Of course they do, up to a manageable point. Beyond that requires the vigilance of a Spanish Inquisition, the taxes of King Louis XIV, and the insensitive heavy hand of a centralized federal bureaucracy under a Stalin."

"Then what's the answer?" called a voice.

Jodi took a deep breath and looked across the sea of faces before answering.

"Seek a harmonious communicating relationship of the goals set forth by our founding fathers. We need not, nor should we, pursue total ideological, ethnic or political universality on any subject. That is the path to unending internecine warfare and diversity."

Jodi started to step away from the podium, then returned for one final observation.

"That's why our forefathers separated church and state--to prevent the mandate of a universal ideology." She paused, adding. "They established separate judicial, legislative and executive branches of government to prevent universality. Trouble is, those branches haven't stayed separate. Each has stuck his nose so far into the other's back yard he thinks he's invited to every barbecue."

38. PRESIDENT UNDER FIRE

Excitement built in the capitol as TV and local newspapers vengefully pumped up the idea of justified impeachment. They harbored resentment over the way President Ericson had snubbed the White House Press Corps. Jodi smiled as she sat behind her desk in the Oval office.

"Brawl promises to be a dilly," she said softly to herself as she read the headlines. "But it's gonna take more than congressional horsepower to pry this filly out of the saddle."

Jodi tossed the offending newspaper in the waste basket and punched the intercom button to her secretary.

"See if General Hardin can meet me after lunch. Say about three o'clock. Also get Harriet Goodwin and Jerry Quinlan. Want the Secretaries of Education and Justice in the same meeting. Thanks."

After a moment's reflection the President punched the intercom again.

"At 5:30 set up a meeting with the heads of DEA, EEOC, EPA, the Superfund Administrator and EDA."

"EDA...President Ericson?" the secretary stammered.

Jodi's face lighted with a broad smile. Another non-entity, even to the President's secretary.

"Econmic Development Administration," laughed Jodi. "A waste ridden agency which loses $1 billion a year. I've got bad news for the lot."

General Hardin was fifteen minutes early. Goodwin and Quinlan arrived at the Oval office several minutes late. Cabinet member holdovers from President Roland, they weren't accustomed to jumping to a White House summons on short notice.

President Ericson got right to the point.

"I expect to make drastic changes in the way the nation handles crime and education," said Jodi, addressing the trio seated in chairs drawn up to face her desk. "Your departments will be particularly involved," she said, speaking directly to Goodwin and Quinlan.

Secretary of Education Harriet Goodwin squirmed uneasily. Her morning had been spent appeasing the National Education Association. She had promised the powerful teachers' union larger outlays for security and more money for social workers and truant officers to persuade kids to return to school.

"We do need help, Madame President, Goodwin said tentatively. "NEA demands better security and..."

President Ericson interrupted.

"Frankly, I don't give a damn about NEA union demands. They care nothing for the students, only for themselves." Jodi picked up a yellow pad, ignoring Harriet Goodwin's thin tightened lips of protest.

"For several days General Hardin and I have been devising a plan. Today I've decided to plunge ahead. Waiting for perfection puts a campaign in never-never land. General Hardin will explain his part of the operation."

The new Chairman of the Joint Chiefs, sitting in a wing-backed chair at the side of the desk, spoke carefully.

"On President Ericson's orders I've established a task force within the Department of Defense to create a Youth Work Corps throughout the country."

"Stealing Roosevelt's CCC thunder?" Attorney General Quinlan smiled. "Nothing illegal about that."

"The camps will be for boys 14 to 18 years of age," said the General.

Harriet Goodwin immediately raised her voice in protest.

"Junior high and high school ages. Kids that age should be in school."

"Right, but they're not," said Jodi. "These youths will be picked up from the streets and from those booted out of school."

"Few students are ever booted out of school today," said an indignant Secretary of Education. "Our schools make every effort to keep kids in class."

"Maybe so, but 8,000 kids were truancy problems in the District of Columbia alone," retorted Jodi. "That's about to change."

The Secretary of Education nervously looked over the top of her wire-rimmed glasses. She and Quinlan were cabinet holdovers. Neither saw eye-to-eye with this fresh young maverick.

"Not keep kids in school, Madame President? Doesn't make sense. Teachers spend half their time making sure kids stay in school or there'd be even more truancy."

"When that time would better be spent in teaching," said Jodi.

General Hardin explained. "President Ericson wants the time in class spent on teaching, not discipline. A few hoodlums disrupt too many classes, particularly in the inner cities. Instead of coddling them, expel them. Let the rest learn in an atmosphere of peace."

"So expelled misfits can roam the streets and cause more mischief, including the creation of more teen-age pregnancies?"

"Absolutely not," said General Hardin. "Schools will have the authority to expel disruptive students. When they do, they'll immediately call the Youth Work Corps and the expelled students will be picked up and sent to camp."

"Prisons?"

"In no way. Anymore than members of the military are prisoners."

Harriet Goodwin pushed back strands of thin gray hair and adjusted her bifocals.

"Our educational system's been fighting an all-out battle against *choice,* now we've got to fight the military against taking more kids from our schools?"

"A better school system wouldn't have fostered a revolt for choice in the first place," said Jodi.

"We're not selective. We take everybody. The NEA's spent enormous effort and money to prevent *choice* from dragging pupils from the system."

"If you wish to fight choice, you'd better do it with quality education, not NEA union muscle. People won't stand for unions running their schools. Neither will I," said the President.

General Hardin attempted to soothe Harriet Goodwin's injured feelings.

"Education will be paramount in Youth Work Camps. School will occupy the full mornings of every day except Sunday. Afternoons will be spent in military discipline exercises and study. Later we may include such worthwhile study activities as cleaning up rivers, fire hazards and other community projects."

"What makes you think problem youths will behave any better in camp than in school?"

General Hardin smiled. "I've known hundreds of difficult kids. I've yet to see one win out against the military. Almost without exception they come to take pride in themselves. They ultimately march in parades to the beat of the drum and hearts of high dignity."

"Eating the same food, wearing the same clothes, doing the same chores abolishes macho envy. Fancy shoes, wrist watches, leather jackets or weird haircuts aren't allowed," said Attorney General Quinlan. "I know. I served six years in the army."

"It's a great leveling and learning field," said President Ericson. "All these youths off the streets will lower crime, eliminate need for more prisons, reduce health care emergency costs by one-half, since most are caused by gun shot wounds, and start to again civilize our society."

"A large order," said Secretary Quinlan. "Opposition from congress will be frightening."

"As of now there is no congress, Mr. Quinlan. My National Emergency took care of that...at least for the time being."

"If it holds up in court!"

"Your duty to see that it does." President Ericson eyed her Attorney General quizzically for a moment. "Think you can handle the job?"

"Oh yes," he said quickly, noting the steely look in the President's eyes. He knew this President would zap him quicker than a fly swatter if he didn't pursue his task with zeal.

"Good," said Jodi. "There is absolutely no way to reach and train the bottom or disadvantaged of our society except through Youth Work Camps."

She explained.

"The IRS alone will require thousands of manual processing jobs for technical people in upgrades extending into a $10 billion effort. And that's only one example of thousands of new jobs which will take place in the U.S. Just as young workers learned new skills for the industrial revolution, they must now prepare for the information revolution. This can't possibly be done by our present educational system."

"Why not?" expressed a cynical Harriet Goodwin.

"Because the industrial revolution took 'hands on' mechanical skills. Today skills are more cerebral than motor. Operators need basic knowledge and understanding."

"May take years in a work camp to get such knowledge," protested the Secretary of Education.

"Youths need added time and maturity to develop values which lead to responsible choices," said Jodi. "Lessons of history are clear. No species on earth has such a slow maturation rate as humans. From birth to independent responsibility takes not weeks or months but years--nearly two decades. If our youths don't mature properly, we've lost civilization."

Jodi turned to face General Hardin.

"Now General Hardin, about Universal Conscription? When will you be ready to accept recruits?"

"You're going ahead with the draft?" gasped the Secretary of Education.

"Not the draft," said an exasperated President. "Universal Conscription is no more akin to the draft than the King-of-Beasts to a hyena. As I've explained many times, the draft excludes, Universal Conscription includes. All serve; the clubfooted, many functionally handicapped and all except major disabilities. Gives everyone a sense of pride and belonging. Also the country later provides Medicare and social security, so citizens can do a bit for their country in return."

General Hardin supported the President.

"In World War II out of 11 million men less than one million saw combat. With today's computers we can assign men according to their capabilities. Clerks, telephone operators, cooks, maintenance men and a host of other occupations, none of which require carrying bayonets."

"Any questions?" The President addressed the Attorney General with a broad smile. "Lots of work for your department, Mr. Quinlan. No longer will bureaucratic zealots have time to get involved in chasing down diaries, pursuing sex scandals or sending people to jail for killing rattlesnakes, Kangaroo rats or certain type mosquitoes."

"We've plenty to do," agreed the Attorney General. "Fighting your impeachment, the National Emergency, new executive orders establishing Universal Conscription and a Youth Work Corps will keep my people burning the midnight oil."

"Good," laughed Jodi. "I like seeing bureaucrats earn their pay."

The President sat back in her swivel chair rotating a pencil between the palms of her hands, looking first at one then the other. They squirmed nervously.

"We face another major problem."

Quinlan's features grimaced, as if he was about to receive a blow to the jaw.

"What's that?"

"Create jobs."

"We'll have taken a big load off the job market with Youth Camps and Military Conscription," said General Hardin.

"True enough," said Jodi. "And big tax cuts will let the economy grow and expand. I'll announce major cuts in a few days. Under National Emergency powers I'll make all cuts and changes by Executive Order, bypassing congress."

"Another headache for your Attorney General," sighed Quinlan.

"Bill me for the extra midnight oil," laughed Jodi. "But I've launched my attack. There will be no turning back. Those who want out better speak up now."

She eyed each of them individually for any sign of response. A long period of silence ensued but there were no takers.

"How in the world can you cut taxes and fund all these programs?" Harriet Goodwin finally exclaimed in a sniff.

"No new prisons, dissolving or merging dozens of agencies, pulling billions in bureaucratic tapeworms out of layers of pork fat will do wonders," said Jodi.

"In the military thousands can be supported for the cost of billions in Labor Department training dead ends," said General Hardin. "Also we can keep up advanced research and development with operable prototypes but at the same time cut back on the enormous production costs of expensive and exotic weapons. That way, hundreds of thousands of soldiers can be educated and trained with no increase in taxes."

President Ericson offered additional ideas.

"There are numerous ways federal funds can be allocated more efficiently: drag billions out of atomic waste programs which have achieved nothing; provide money only to companies who keep administrative costs on government contracts below specified percentages; suspend the high cost of the Davis-Bacon law; limit money to Universities whose faculties teach students less than 20 hours per week. Finally, adopt my amendment for reform."

"A large order," said Quinlan, scratching his head with a forlorn glance through the windows into the Rose Garden. He saw his planned vacation for scuba diving off Belize dashed with cold water.

"By the time we've finished I want the inner cities to look like ghost towns for crime at night," said the President. "Youths will be at home or in camps studying. Later, together with the military, they can help clean up rivers, work in national parks, engage in rebuilding the nation's infrastructure of roads, bridges and the abandoned inner-city wasteland."

"Requiring skills they don't have," protested Harriet Goodwin.

"And which our schools are not providing," said Jodi. "An educational system which stops educating students to be American has failed its most essential function."

"I don't follow?" said a perplexed Harriet Goodwin.

"Bilingual education. Subsidized to the tune of $260 million annually. You now graduate students unable to read their diplomas in English, yet are able to get a driver's license, and vote in elections where they can't understand a word candidates speak.

"For your unionized educators, bilingualism has become a $10-billion jobs program. But when youths reach the workplace, no one wants them. Employers call 20 percent lousy. One-third of surveyed bosses declared 25 percent of new graduate employees incompetent. Many employers called them uneducated, untalented bums who drain the profits of any business unlucky enough to hire them."

"Bigots," cried the furious Secretary of Education.

"I hardly think so," said Jodi. "They're just fed up."

"In the Youth Work Camps youths can learn English and acquire many skills for later jobs," said General Hardin.

"Today, machines are only tools," said Jodi. "Under proper military supervision youths will not be required to perform heavy work beyond their ability. Instead they'll learn. The military throughout history has furnished excellent teachers. Computers, vocations of all kinds are available. There's a thousand things our youth can learn besides crime. We need only to get their attention."

"Like the Missouri farmer getting the mule's attention by whacking it with a two-by-four," said a skeptical Secretary of Education. "Brutal."

"Perhaps," said Jodi. "If that's what it takes to turn this society around. Right now people are fed up with the present state of the nation. More money for incompetent programs would be better spent on raising sheep--something to feed those who will never assume responsibility for themselves."

* * *

AT 5:30 five unhappy people confronted President Jodi Ericson in the Cabinet Room. In their uneasy state of minds they were in no mood to take in the beauty of the scenery as a slight breeze wafted through the partly opened windows.

Coffee was available on a portable sidebar but only EPA Administrator Howard Bloomberg helped himself before all were seated around the mahogany pedestal conference table purchased by President Nixon.

After a few preliminary social remarks, President Ericson got down to business.

"I've summoned you five, the first of many, in order to initiate a complete change in the way government responds to the people who pay their wages. I think too many of us forget that connection."

She turned to the Economic Development Administrator, Jerry Nussbaum.

"I'm starting with EDA because it offers the easiest solution." Jodi took a deep breath and exhaled slowly. Her tasks weren't going to be agreeable.

"Jerry, I want you to dissolve EDA completely. It has become useless as well as expensive." At Nussbaum's look of terrified consternation, she added gently. "It need not be done overnight. You have twelve months. Continue to honor all contracts in force but begin to trim at once."

He lifted pain-teared eyes without comment.

"You understand?" Jodi insisted.

Jerry Nussbaum nodded his assent.

President Ericson next directed her attention to Superfund Administrator Fred Lazer.

"Regretfully, the American public knows the Superfund has been the most expensive boondoggle in our history. Even exceeding the bank scandals where many went to jail."

Lazer face turned red.

"I hope you're not accusing..."

"I'm accusing no one, Mr. Lazer. We've all been at fault. None more than the Executive Branch and Congress. But once we find we're in quicksand we should get out as best we can."

"How?"

"Not another dime is to go to lawyers. Nearly $50 billion has been squandered in the 14 years since the Superfund law was enacted and only 50 sites--out of 1,319 established priority sites--have been cleaned up. But billions have provided a steady income to thousands of lawyers and consultants."

"It couldn't be helped," Lazer said defensively. "The laws generated thousands of lawsuits."

"From now on all agencies under my control will take another tack," said Jodi.

"And what tack's that?" defensively queried EEOC administration Elias Brown.

"Actions legally taken years ago will not be retroactively penalized," Jodi said grimly through gritted teeth. "I intend to fire anyone I hear of doing so." She directed a meaningful glance at EPA's Howard Bloomberg who, through her own administrative oversight, was still on the job. He kept his eyes on the coffee cup.

"A position I've long advocated but for which I got little support," said Lazer.

"Well, you've got mine now. I want to see Superfund spending cut by two-thirds in the next twelve months. And retroactive penalties in all agencies are henceforth forbidden. Our job is to correct, not punish. It's vindictive to punish our own people who at the time did not realize they were causing damage when there were no legal restrictions in place. You are no more entitled to beat the hell out of innocent business than innocent children. Mental midget Rhodes Scholar Reich tried that approach and it failed. Our job is to improve, not excavate for dead bodies. The U.S. is not a penal camp to be raided by jailer regulators."

"We've got a lot of zealous regulators out there," said Lazer.

"Fanatics. Not regulators," said Jodi. "Congress doesn't seem to get the message that a major reason for their unpopularity is the laws they've passed permitting agencies to destroy lives. Zealots on the lunatic fringe have taken law enforcement into their own hands. Nightmare depredations by bureaucrats would fill a book. If congress can't correct their own predatory laws, they should shoulder the blame and get booted."

The Reverend Elias Brown, Equal Employment Opportunity Commissioner, had retained his job through three Presidents. He felt secure as a buddy of the powerful black caucus.

"Certain employers must be penalized for past wrongs," he said, speaking out defiantly. "I intend to keep penalizing those who disobeyed the 1964 Civil Rights Act."

"You've stood the Civil Rights Act on its ear," said Jodi testily. "In too many cases you've turned an innocent act into a federal crime."

"Not so, President Ericson. An entirely false statement," Brown said, raising a voice to a high pitch.

"Companies were forced to hire criminals ahead of ordinary citizens because EEOC declared that "discrimination against former convicts had a 'disparate impact' on preferred minorities,'" said Jodi. "You even had employment tests 'racially normed' so blacks could beat out whites who had higher test scores."

"Corrections were needed," Rev. Elias Brown stubbornly insisted.

"You ruled Chicago's Daniel Lamp Co. employed too many Hispanics and demanded a $133,000 penalty for not employing three more blacks," retorted Jodi. "What nonsense. Your agency's little better than Nazi SS troops foraging for victims."

"With all due respect, you're the one preaching nonsense," said Brown, stiffening in his chair.

President Ericson surveyed the room quietly for a full minute. She glanced out the window at the gardener peacefully trimming the hedges and tending the flower garden. Some people don't know how lucky they are, she fretted. Finally, she turned her attention back to Rev. Elias Brown.

"Reverend Brown," Jodi said quietly, "I think you should adopt new procedures."

"What's that?"

"Thousands of EEOC employees are on staffs of most agencies, right?"

"Practically all agencies. We have a huge impact."

Jodi looked squarely at Elias Brown. Her eyes were cold-blue steel.

"I want you to leave one representative in each agency and fire all others. If any agency needs your help you can furnish help from your headquarters."

Reverend Brown glared back at President Ericson with mounting fury.

"President Ericson, you're crazy."

"Perhaps. I'm also cutting your budget by two-thirds. That should eliminate the wild-eyed zealots on the fringes who wantonly throw hand grenades into people's lives."

"I'll do no such thing. I plan to take this up with the black caucus."

"Fine. The legislative branch is across town. Meanwhile, I expect to have your resignation on my secretary's desk as you leave."

The others at the table sat paralyzed as they watched Reverend Elias Brown storm out of the Cabinet Room and slam the door behind him.

President Ericson smiled sweetly at the group after he left.

"There are those who disagree with their President. It is their right," she said. "Also it is the executive's right to choose agency leaders." Jodi hesitated momentarily.

"Now where were we? Oh, yes," she answered her own question.

"I hate to present such a negative front, but spending must be brought under control and any diet is not only uncomfortable but downright painful at times."

Jodi smiled at DEA chief Calvin Manson. He reacted as if standing before a firing squad.

"I suppose you know what's coming?"

"A serious depletion in my budget?"

"No."

"No? Then what?"

"The same job I gave Jerry Nussbaum."

"The complete elimination of the Drug Enforcement Administration?" said a dismayed Manson.

"Yes. The agency takes property without due process, spends millions on obsolete helicopters and interdiction hardware, which is useless against speedier drug craft. Anyhow, I intend to beef up the defense department to take over the job which your people have turned into a greedy grasping organization of self-anointed czars outside the law. In too many cases they forgot *due process*."

Calvin Manson, stunned, asked.

"Am I fired too?"

"Not at all," said Jodi pleasantly. "Unless you object to dismantling the organization completely. Do you accept the mandate?"

Calvin Manson gulped, his Adam apple bouncing up and down before he managed to blurt.

"Of course, President Ericson," he said, glancing at Jodi's expressionless face. Her steely visage assured him he had the same choice Hitler gave Poland.

Jodi said. "For your information you people are not the only ones singled out for spending cuts. I have on my desk orders to combine the Departments of Commerce and Labor; abolish the Departments of Education, Energy, Transportation, and Housing and Urban Development. Some functions will be transferred but most already are provided by the states. Regulatory agencies stretch as far as the eye can see, each with unsupervised zealots with clubs in their hands to punish imagined criminals--property owners and businessmen.

"As columnist Haley Barbour put it: *The American people don't want government to promise a 'chicken in every pot' because they know if it comes from the government it'll cost five times more. And by the time it clears FDA, USDA, EPA, OSHA, ICC, FTC, and SEC, it'll be spoiled."*

Manson nodded agreement. A wise decision on his part.

"Jodi looked at her watch and then at Howard Bloomberg, who had never opened his mouth throughout the entire proceedings. He remembered too vividly their initial luncheon when he had tasted her tactics as a U.S. senator.

"Mr. Bloomberg," Jodi said in soft tones, "I'd like you to remain after the others leave. We've many problems within your agency. EPA, with such a huge portion of the federal budget, needs mammoth refocusing. Details and goals have gone completely askew. There's no need to involve others."

Jodi turned to the silent group.

"This meeting's over. I thank you all. I've no doubt you'll deliver as promised. I'll keep watch."

With those last words Jodi abruptly resumed her seat and observed all but Howard Bloomberg silently depart the room like numbed defendants just handed death sentences.

Howard Bloomberg shifted his weight with the queasy feeling of a victim on the scaffold. Had she left him with a rope around his neck on purpose? When would she spring the trap door?

The arrogance Bloomberg displayed within his own agency was gone. He spoke in pleading tones.

"I know you plan to fire me, President Ericson. But couldn't it wait until all the hub-bub surrounding impeachment dies down? My agency's in a state of chaos?"

"You think I plan to discharge you because I harbor some deep resentment?"

"That and because you don't agree with EPA's philosophy."

"Mr. Bloomberg, people want clean air and clean water. I lived on a farm. Farmers want the barns cleaned, silos and granaries fumigated, wells sanitized and debris removed from our streams and rivers. They don't want you to burn down the barns, destroy the silos, fill in the wells and declare non-flowing mosquito beds as wetlands."

"Actions which need to be done in many cases to protect the environment," said Bloomberg stubbornly.

"Also which need to involve a bit of common sense. You don't seem to grasp what every child of ten knew years ago--you don't kill the goose that lays the golden egg."

"The state of our environment requires drastic action."

"We all want clean air and clean water objectives achieved," said Jodi between tight lips. "But we don't want the job done by fanatics who resort to coercion. Hauling Spanish Inquisition methods to the new world was not the intention of our forefathers."

"Sometimes the public needs to take a few aspirins to rid a headache," said Bloomberg.

"Right. But forcing a bottle of aspirins down their throats will kill. Your organization's become little better than cult zealots. Total destruction of anyone who doesn't agree with each and every action."

Bloomberg saw his case was hopeless. There was no doubt the President intended to remove him. He had nothing further to lose.

"You're firing me because you dislike me. A president should not stoop to such conduct. You're no better than the zealots you decry."

Jodi bit her tongue to keep her temper under control.

"No, Mr. Bloomberg. That's not the reason, though I'll admit I've come to dislike you. We came to dislike Hitler not because he wore a mustache, was an Austrian or didn't belong to the aristocracy. He was disliked because his actions became hideous."

"The EPA's only responded to congressional mandates," said Bloomberg stiffly.

"EPA fanatics seized land and property without due process: acts in which you've acted as prosecutor, judge and jury. Then if you couldn't seize property you made sure the rightful owner can't use it either."

"Congress has voiced no objection to the way we've operated."

"Sound judgment is an expected responsibility of the head of an agency as powerful as the Environmental Protection Administration."

"Why keep me here merely to fire me?"

"I wanted no false rumors or misunderstandings. You're being relieved because you couldn't curb your appetite--exercise a bit of common sense--not because the objectives of EPA are wrong."

Bloomberg stood up, his eyes flashing hatred.

"I'll be back in this job once you're impeached, Madame President," he stormed, striding for the door.

39. SPEAKER LOSES COOL

Speaker 'Fat Albert' Metzer was in such rage his whole body quivered in rolling undulations. He stood next to Senator Mordecai Meyer and Reverend Jesse Rowan in one corner of the Carlton Hotel's first floor conference room, flanked by a garden patio with partially opened glass doors. It was a warm early September day.

The room buzzed with highly animated conversation concerning the possibility of impeaching President Jodi Ericson. Eleven senators and forty-seven representatives were present, representing the bulk of those who had returned to homes in Washington. Not all were of one mind, some attending without invitation.

They were about to engage in a luncheon meeting coordinated by the Speaker, Senator Meyer and Chairman of the House Judiciary Committee, Congressman Rowan.

"That bitch each day becomes more dangerous to congress than Hitler to the Jews," fumed the Speaker.

"She exudes such innocent beauty, she's hard to attack," said Rowan. "Like slandering the leader of my church choir."

"Her own hatchet man, Tony Pellini said it best," said Senator Meyer.

"How's that?"

"I overheard the remark, meant as a joke."

"What'd he say?"

"Jodi's flawless on the outside; on the inside she's loaded with demons."

"Well, she's pulled out the demons for us," complained Rowan.

"You've still got the power of impeachment," said Mordecai with an unhappy grimace. "Best get on with using it before she takes that away too." He glanced at the members beginning to take their seats at the luncheon tables, arranged from the Speaker's podium in the form of an inverted 'U'.

"Time to take our places."

After lunch and several martinis Speaker Metzer stood up and rang a glass with his spoon. He was feeling mellow and confident.

"We're here to do the nation a service," he said without preamble. "Rid the country of another Madam La Farge, who's determined to see members of congress guillotined."

"Here...here," shouted Reverend Rowan, forgetting for the moment he was not leading his congregation.

The Speaker ignored the rumble of disapproval from the far left. He hadn't invited the small group of four dissidents anyhow. Senator May Trudeau, Nevada; Congressmen Jack Pace, Nebraska; Charles Irwin, Wyoming; and Joe Bass of Florida were all friends of the President.

"Sorry we can't meet in our own building," said the Speaker. "That she-monster's replaced our Capitol Hill police with the National Guard. She's got them placed inside all entrances, the underground garages and manning the corridors."

"But not inside our offices," cried Congressman Bass.

"We're free to visit them if necessary," said Congressman Pace. "I just came from mine."

"Sure," said the Speaker in scathing tones. "Ten Senators and twenty-five Representatives at a time. Like children being let on a bus."

"The President wants to prevent us settling in and causing trouble like Yeltsin's original Parliament," said Senator Trudeau, smiling sweetly at the angry Speaker.

"She's cut off all phone communications," Mordecai sputtered impotently, "so what good's an office?"

Senator May Trudeau laughed heartily, annoying the Speaker with her interruption.

"Congress without a mouthpiece is like a chimpanzee without arms. Frankly, I like the opportunity to get some work done without the damn telephone ringing off the hook."

Chairman Jesse Rowan pulled a yellow pad from his briefcase.

"Let's work on articles of impeachment," he said, fervently resolved to get ahead with the downfall of an enemy.

"Great idea, Mr. Chairman," said the Speaker. "Take over. Impeachment's your ball game." He resumed his seat as the Chairman of the House Judiciary Committee approached the microphone.

"I've got a list of sins the President's committed which deserve impeachment," Rowan said, placing his notes on the podium.

"I've already drawn up three. Ought to be enough," grumbled Mordecai.

"We'll undoubtedly use those," said Rowan, accommodating the Senator. "But a list of her arrogant usurpation is in order for the media and the public. Will strengthen our hand."

"Of course," said the Speaker. "Proceed."

"The President has created, under the military which is difficult for congress to touch, a Youth Work Corps taking youngsters from junior high and high school."

"Obviously an invasion of the home," injected Mordecai.

"Without congressional approval she established Universal Conscription."

"Which the courts will knock down," said the Speaker.

Senator Trudeau commented to her colleagues so everyone could hear.

"Camps will be up and running long before our courts take any decisive action. General Hardin opened many this past three weeks."

Reverend Rowan paused in his presentation to nod his head in May Trudeau's direction.

"Our esteemed Senator exposes another reason for impeachment. The President's got the Attorney General scared stiff. He's setting up more hurdles for congress than an Olympic track meet. Look how the President's damaged education."

"Damn hard to damage that." Congressman Joe Bass was a strong advocate of choice.

Chairman Rowan continued after an impatient glance at the Texan.

"Our own President tells teachers throughout the land to expel young students, exactly opposite our policy of trying to keep kids in school."

"The Youth Corps will pick up truants and expelled misfits," cried Bass. "Not a bad idea if you don't permit choice."

"Unconstitutional as hell," shouted the Speaker.

"Doesn't matter long as she's got the military behind her," said Congressman Bass. "Joint Chief's working hand-in-glove with the President."

Chairman Rowan looked again to his notes. He didn't appreciate the flak he was getting from the other end of the room.

"President Ericson's moved without congressional approval to cut the hell out of our whole system of bureaucracy," he said wrathfully. "She's ordered the State Department to cut in half all expenses of overseas embassies and consuls. She's issued orders to abolish the DEA, the Departments of Energy and Education, merge Labor and Commerce, and tear the guts out of more than 100 other agencies."

"Good," said Senator Trudeau. "They're complete waste of money."

"Our dear President," continued the undaunted Reverend Rowan, "just yesterday replaced the head of EPA..."

"Hopefully to bring in a real scientist with a smatter of common sense," hollered Bass.

"We've got enough ammunition to impeach Napoleon," called out Senator Meyer in an effort to support the beleaguered Rowan.

"Who's to fire the guns?" Senator Trudeau laughed in her soft lisping voice. "Politicians didn't have much luck in firing Napoleon."

Congressman Rowan was not finished.

"The President's told General Hardin to lend military support to Immigration, not only to stop the flow but return illegal immigrants."

"Merely implementation of California Proposition 187," said Bass. "Something the voters wanted. Despite judges acting outside their jurisdictions, President Ericson has supported the will of the people."

"An irresponsible task, creating enormous hardships," retorted Rowan.

"Hardships on who?" shouted Bass. He'd been overloaded with problems in his district. "American taxpayers have been paying through the nose for added costs to welfare, food stamps, emergency health care, schools and 100 other benefits for illegals while our own unemployed go begging."

"Using the military against unarmed civilians is obscene," said Rowan.

"You call out the military when unarmed civilians resort to rioting and looting," said Bass. "Today's coastal border situation is intolerable for a soverign nation."

"Nonsense."

"Governor Wilson was on the front line in California. He said we needed the skills of the military. There are more law enforcement personnel guarding the U.S. Capitol building than were protecting our sovereignty. Shows we're not much better than the cowardly congress which fled before the British. We think of our own skins first."

Rowan flashed Congressman Bass an unhappy look. He didn't want to tackle that subject. Instead he returned to his assault on the President.

"Our new Iron Lady orders the bureaucracy cut in half in the next five years," said Rowan. "What's she got against the bureaucracy? She determined to rip out its guts?"

"Not its guts," laughed Senator Trudeau. "Its blubber. It's filled with enough bloated calories to feed the homeless for a generation."

Chairman Rowan raised his hand for peace.

"I could go on and on about this President's unconstitutional acts, but I've outlined only the high spots. President Ericson's many acts deserve impeachment. Our absent colleagues will approve. I've contacted nearly 100 and got only a few dissents."

"Championing your cause, no doubt?" Senator Trudeau said with a skeptical tilt of the eyebrows.

"I could introduce more," snapped Reverend Rowan. "I've got a list..."

"Leaked by a bureaucrat?"

"Of course. Matters like this should be made public."

"Which you plan to leak to the press?"

"Naturally," smiled Rowan. "Soon as I leave this meeting."

"Will have little impact," said Senator Trudeau. "President Ericson informed me she goes on national TV Sunday, two days from now. At that time she'll rally the people."

Retorted Reverend Rowan.

"By then she'll be rallying a horse long dead and buried."

"The people will have their say."

Reverend Rowan grinned.

"Like rallying the spectators at a football game. Great noise-makers. But they're not on the field of play. Only congress impeaches and convicts. Hitler has a better chance for a monument in Israel than Jodi Ericson escaping impeachment."

40. OCTOBER: IMPEACHMENT LOOMS

Two blocks from the White House in the ballroom of the redecorated Willard Hotel Speaker Metzer and Congressman Rowan began building the case for impeachment. They molded their procedure on Nixon's ordeal.

Reverend Rowan, Chairman of the House Judicial Committee responsible for impeachment hearings, took charge. He notified members of his committee, most still in recess back in their districts, that hearings would begin the moment they returned. Meanwhile they were to hit the media and their constituents about the gross misbehavior of President Jodi Ericson.

The Speaker released to the press the final articles of impeachment--subject to amendment on the return of members. Again, they were practically identical to those used against President Nixon.

1. Obstruction of Justice
2. Abuse of Power
3. Contempt of Congress

By now all parties knew impeachment was an emotional rather than a factual undertaking.

The White House Press Corps, shut out of the President's loop, responded by gleefully jumping on the Speaker's bandwagon. Impeachment was a much more exciting news story than any constitutional amendment to reform congress.

Events from this point began taking an ominous turn for President Ericson. The black caucus solidly backed the vengeful Rowan, carrying the bitter defeat on his Vice President nomination like a chip on his shoulder. The powerful teacher's NEA union marshaled forces against Jodi. With all the resources at their command, the bureaucracy in Washington connived with belligerent regional offices to leak and distort information.

Instead of pushing through a controversial amendment Jodi found herself fighting for her very life. The battle line was clearly etched in the sand. President Jodi Ericson vs the taxers and spenders. The Executive against an Imperial Congress.

Jodi believed congress was like King Louis XIV. His 70 years of taxation laid the groundwork which eventually destroyed France. His son followed his lead in creating a French revolution which became inevitable. Napoleon's rise was thoroughly predictable. He took the rabble rousers and unemployed young off the streets and put them in the military to fight foreign wars. Always a popular way to divert attention from domestic woes.

Powerful leaders know when the idle are left to roam the streets every possible evil springs from the chaos. Crime, delinquency, teen-age pregnancies, unmanageable schools and a permanent beggar--now called welfare--class.

Instead of calling out the guard to wage war against internal dissidents, rulers like Napoleon and Hitler put the rebellious into the military to wage war against evil foreign devils. Instead of tyrants, they became heros.

President Jodi Ericson, ruefully going over her options with Tony Pellini in the Oval office, determined to launch a full counter-attack on congress. There was no other way out.

"To win a war you've got to first fight the battles," she said, setting the coffee cup down so hard it chipped the saucer.

"No question," agreed Tony, absently gazing through the windows at the early morning haze in the Rose Garden.

Brushing her hair off her shoulders, Jodi smiled across the desk at Tony.

"When Yeltsin took up the gauntlet thrown down by a belligerent Parliament he knew his first job was to drive out the old hard liners," she said. "In later elections he made the mistake of putting himself above the fray, disdaining to lead his troops and fight the battle. A bad mistake which almost cost him his job."

Tony had yet to adjust to this dynamic female as President of the United States. Just when she appeared most kittenish, she was likely to turn into a jungle tiger.

Tony said. "Yeltsin may have learned from Nixon. When Nixon tried to carry on, ignoring the proceedings against him, he was driven out of office. When parliament went too far, Yeltsin stepped in--even brutally."

"My point exactly," said Jodi. "Nixon failed to battle for his own survival--fight back, destroy the tapes, tell his foes to go to hell. I plan to do battle according to my father's favorite: 'It's not the size of the dog in the fight, but the size of the fight in the dog, that counts.'"

41. OCTOBER 30: ON THE HUSTINGS

Attorney General Quinlan warned President Ericson as the group sat facing her in the Oval office.

"You've barely sixty days until members return demanding full access to their offices. They're rapidly building alliances with the media. Unless you somehow counter the tide of revolt, impeachment seems a certainty."

"You've roused vengeful giants in the bureaucracy, the NEA teachers union and the legal profession," said Harriet Goodwin. The Secretary of Education wasn't happy in her role of downsizing and eventually eliminating her own department.

The media thinks I'm a bigger patsy for impeachment than Nixon, Jodi thought grimly. They sense another Watergate. Today's media imagines themselves investigative sleuths, rather than honest professionals reporting the news. They'd rather report a scoop from an unauthorized source, no matter how big the lie, than laboriously spade for the truth. Jodi promised herself never to let down her guard before the media.

Tony Pellini voiced a note of encouragement.

"You hit the road for a solid two weeks, starting tomorrow. Stops are among friendly down-home people. The kind you were so comfortable with in Governor Aker's campaign."

Jodi smiled, biting off a torn fingernail.

"I knew I could rely on you to put me to work. Where's our first stop?"

Jodi had called Tony in from Nevada to arrange her schedule. She trusted none of her cabinet officers, inherited from the past president. She kept most because she hadn't wanted to make waves following his assassination. Not that they were disloyal, only in the coming crises she needed brutally courageous troops with a zealous track record on her behalf. Only Tony had been with her through past wars.

"We start in California and work our way back east." Tony grinned as he referred to his yellow pad. "Get away from Washington scalpers until you've got a few solid speeches under your belt."

"Concerned about my self-confidence?"

"Broadway tests plays with shake-down runs," said Tony. "In San Franciso you'll face an audience plenty tough enough."

President Ericson turned to her Attorney General.

"You've got people in the Speaker's camp. What're they up to?"

"Organizing a cadre to face you down at every opportunity," said Quinlan. "Remember the old time debates? Whenever the President spoke on TV, a team of antagonists came on afterward to refute everything he said?"

"I read about them," said Jodi. "Speaker's got the same idea? Setting up a road block at every intersection?"

"Afraid so. They're pulling out all stops, using their local congressmen to get on every possible platform with you."

"Regular debate at each site?"

"Heckling. Or guerrilla warfare. Anything to prevent you from making an uncontested speech."

President Ericson silently chewed her lower lip for a full minute. A habit Tony recognized when she was under stress.

"Perhaps their ploy could boomerang," Jodi said slowly, pinching her forehead in concentration. "Genghis Khan always took the offensive, though invariably outnumbered. My opponents will be defending indefensible tax and spend positions. The American people want coherent, workable solutions."

"Which takes us back to fundamentals," said Tony. "In football a player better learn to tackle before he tries to stop an open-field runner. Congress has avoided tackling fundamental flaws by waving bribes at every punt returner. Suddenly, they're finding themselves out of bribe money and they're at a loss. They never learned how to tackle basic problems."

President Ericson got up from her chair to gaze out the windows. A habit when confronting a stressful situation. The others, sitting silently, waited patiently.

Jodi spun about, her features dimpled in a big smile.

"You've rung the bell, Tony. We'll do what you accomplished for me in Nebraska. I'll invite the Speaker's hecklers up for debate. They're programmed by the Speaker. Away from Washington they'll be fish out of water."

"Mentally, they were beached when they started to work for the Speaker," said Tony, referring to Jodi's long history of troubles with Metzer's office.

The Secretary of Education smiled serenely, her message was clear. Keep your hands out of our agencies. Instead she said in a voice of apparent concern.

"Why bother yourself with all these details, Madame President? Let us agencies handle the work."

"Because it's the details which determine the quality of the product," said Jodi. "Something our automobile executives forgot until the Japanese stole their markets. Paying attention to details is also the only way to improve the quality of government. Too many wave their arms and promise change, then promptly forget the details."

Harriet Goodwin's face turned red at the subtle reprimand.

"Educators can't be turned on and off like spigots," she murmured resentfully, casting her eyes to the floor.

"They're unionized," said Jodi. "The NEA handles them like robot spigots. I'm sure they can adapt to a bit of quality control in their spigots."

Jodi returned to her desk and pulled out a mammoth book on federal departments and agencies. Thumbing through the massive document she turned to her Attorney General with a wry smile.

"What have I got here?"

A puzzled Jerry Quinlan answered. "The Federal Register."

"Correct. Now I've reached OMB where it lists one man assigned to each agency to hear its budget problems."

"Yes."

The Attorney General wondered why the President was playing games with him. She often did that before springing a trap. A favorite ploy of hers. The document's name was inscribed in large letters on its cover.

"So what's wrong with assigning one lawyer from the Justice Department to each agency?" Jodi asked.

After a moment's hesitation, Quinlan said. "Nothing...nothing at all. We have people..."

The President interrupted. "Now I want you to do the same with lawyers...assign one from a pool in your department to each agency."

"That should be no problem, Madame President. We already have considerable liaison..."

President Ericson continued.

"Now listen carefully. Your next step is to fire all lawyers in each and every agency of the federal government. If any agency needs legal advice they can get it from your pool of lawyers--like any stenographic pool. Lawyers are not sacrosanct beings."

"Fire all those lawyers?" Quinlan was aghast.

"Immediately. Without delay," Jodi said. "Grab only a necessary few from each agency and put them in your department. Unnecessary lawyers--costing taxpayers billions--long have been a source of needless litigation and expense."

"There are thousands..." Quinlan stuttered in protest.

"Which will save taxpayers millions," said Jodi. "Lawyers have metastasized like cancer throughout government. They've become as pernicious."

The Attorney General opened his mouth to express his disapproval when he caught a glimpse of Jodi's grim resolve.

"Yes, Madame President," he said.

Noting his hesitation, Jodi questioned.

"Any reservations about executing my orders?"

Jerry Quinlan knew at once his job was on the line. This iron lady dismissed disloyalty with the aplomb of a Genghis Khan. Her Mongol ancestry was common knowledge. He idly wondered if she were a direct descendant. This was no time to find out.

"No, Madame President. I'll forward your executive order first thing in the morning."

"Fine. The sooner the better." Jodi said. "Then you can follow up with several more fiats."

"Yes, Madame President," said a resigned Attorney General.

"This involves a reform long overdue. The government henceforth will pay compensatory damages when necessary but no longer punitive damages. And lawyers who sue any departments of government shall receive fees no higher than 15 percent of the award." Jodi smiled engagingly at the distressed Quinlan.

"If they object too strenuously I may insist we treat lawyers just like our lawyer congress advocates treating doctors. Socialize the legal profession and require attorneys to work for fees established by a national board."

Attorney Quinlan's face turned purple as he fought to catch his breath.

"A flagrant violation of the constitution."

"Perhaps. But it's also a violation of the equal protection laws of the 14th Amendment for congress to socialize one profession--medicine--without socializing others. A step into a deep swamp."

Quinlan choked out an agonized response.

"The Trial Lawyers Association would challenge us at every step."

"Exactly the reason medical costs have zoomed out of control," said Jodi. "Malpractice insurance. A young doctor has to come up with $50,000 to $100,000 front money for malpractice insurance before he opens office. No doctor just out of medical school can afford such monstrous fees. Older doctors must stick it to their patients. As usual congress holds out hands for their own fat legal perks while blinded to the real causes for high costs."

"Legal fees include considerable research," protested Quinlan.

"Knowledge is why lawyers are hired in the first place," retorted Jodi. "Imagine taking your car to a mechanic who's never worked on an engine. Or a carpenter who has no tools. Or an electrician who doesn't know direct from alternating current. Then wants to charge you for research. They're expected to know their trade. Find and fix problems according to past experience."

"Cutting cost is hard to come by in the legal profession," said Quinlan. "Action is so vague."

"The English language is not vague except when idiots make it so. Cost cutting can apply to motions; to case briefs, research memoranda, strategy reports and thousands of other repetitive documents that flourish like loco weeds in American litigation."

"I'll admit, Madame President, the profession is ripe for automation inspired reengineering," conceded the Attorney General. "But big hitters from Harvard or Yale are hard to change."

"Go elsewhere for *new* blood. Bright youths spring from all sources. Bill Gates dropped out of dull schools. Boiler plate--about 90 percent of each legal brief--is already computerized." Jodi paused, looking Quinlan squarely in the eyes until his gaze dropped.

"A tough sell. Lawyers hold most of the weapons."

Said Jodi. "Today's legal system costs U.S. business $130 billion each year--a cost no other country bears in competing on the global market."

“The judiciary is partly to blame,” protested Quinlan. “It strikes down efforts to control costs on welfare, immigration, education and environmental excesses.”

“The judiciary is merely an older generation of lawyers. Let’s correct zany actions before they become judges. My amendment would limit terms of judges.”

“Getting a handle on the judicial system in this country’s a near impossibility.”

“Got to be worth a try,” said Jodi. “Lawyer judges, with not a whit of sense or feeling, impose retroactive punishments and fines. The next despot will not be a Czar or Dictator but a spider web of lawyers, judges and poisonous regulators.” She paused. “We’re damn near in that unhappy state now.”

Jodi turned to Tony Pellini, dismissing the subject.

“We’re going to make at least a few members happy, Tony. I want you to work with General Hardin.”

“How’s that?”

“Speed up reactivating old military bases still under federal control and rehabilitate them to accept youths for the Work Camps. They’re filling faster than we ever dreamed.”

“Any particular suggestions for General Hardin?”

“I’ve discussed this program with the General at length,” said Jodi. “But there are some priorities?”

“Which are?”

“More emphasis on running the bases with veterans who were terminated by past budget crunches. They can swing into action quicker in providing logistic and instructor support. They’re familiar with the system...and thousands of good men are available.”

“Most sites are in excellent shape today,” said Tony. “General Hardin said maintenance has been provided all these years by the military. Legal and legislative hassles slowed turnovers.”

“A delay that’s proved a boon to us.” Jodi looked over to her two Cabinet officers.

“No need for new prisons when we get all these youths off the street. Many military sites are luxurious. Fine quarters, excellent buildings for school study halls, and recreation facilities in 90 percent of the cases provide a far better environment than the youths enjoy at home."

Harriet Goodwin stubbornly insisted.

"Education's still the name of the game."

"General Hardin's introduced educational tools on a scale impossible outside the Defense Department," said Tony, patiently attempting to smooth the Secretary's ruffled feathers.

"Gimmicks," said Goodwin scornfully. Suddenly her worst fears were being realized. Education in America was fast slipping away from the inbred professional teaching profession."

"Not so," said Tony. "The military's equipping each student with a computer. And the bases have the electrical backup to operate them. They won't stand idle for want of power, like many in school rooms across the country."

The Secretary of Education offered on parting shot.

"A prescription for chaos. Elementary teachers aren't prepared to introduce machinery or assembly line mechanics into school rooms."

"Nonsense," was Jodi's tart response. "Too many curriculums today are shaped by the interests of faculty, not by the interests of the students. If tomorrow's graduates are to compete on the global market they'd best exploit all our resources."

"About visiting?"

General Hardin spoke up.

"Parents, relatives or friends will be encouraged to visit on weekends. There will be no secrets. Only proud displays of accomplishment. But make no mistake. Bad boys will work until blisters on their hands change their minds."

42. NOVEMBER: SAN FRANCISCO

President Jodi Ericson was in San Francisco for her first speech since the Speaker's cadre drew up Articles of Impeachment. Tony Pellini had selected the Presidio Military Reservation on the Pacific for the President's western initiative. It was a huge complex on a point separating the Pacific Ocean and San Francisco Bay. Supporting the southern entrance to the Golden Gate Bridge, the area was a national shrine to Californians.

Although congress had gutted the scenic spot of military operations, the remaining compound supported a variety of activities. Tony selected an immense building for President Ericson's appearance. TV and newsprint had widely publicized Jodi's impeachment confrontation with congress so she faced a packed assembly on the evening of her speech. Loud speakers were erected outside the building to take care of the overflow.

Jodi's talk started serenely enough. She did not mention the attempts at impeachment. She wanted to focus on the rights citizens had let slip away. Persons in the west had been particularly abused by the Environmental Protection Agency, the Drug Enforcement Administration and stepped up IRS seizures.

Following cursory introductions by the Mayor and Governor, President Ericson took the podium. She had insisted on her usual forum which allowed questions from the audience. Red-shirted ushers with microphones paraded the aisles. She wanted to become one with the people.

"Individuals have certain inalienable rights as set forth by the constitution," Jodi said in answer to a question regarding the increase in seizures by the EPA.

"The EPA has no rights, the DEA has no rights nor does the IRS have rights which conflict with your individual rights against the taking of property without due process," said Jodi. "Congress too often has tried to usurp your individual rights by passing laws which patronize your intelligence."

"The courts don't help us either," called a bitter voice.

Jodi acknowledged the complaint with a nod.

"Spanish Inquisition judges also sought to straighten out the wayward minds of citizens.

"History is full of instances where tyrants--whether judges, dictators, Robespierre legislators, Roman senators or parliaments--thought themselves superior intellects to the people. So they took away rights, always claiming such action was for the people's own good."

A man on the aisle spoke into a microphone.

"Hitler took away the rights of the Jews supposedly for the good of Germany."

President Ericson answered.

"Individuals lost their importance. Just as Cromwell knew what was best for Englishmen, he slaughtered the immoral--according to his thinking.

"Every single congress in the past two decades has grabbed more of your hard-earned money, the product of your own sweat and toil. Each time they stick their hands in your pocket it's for some bureaucracy--for your own betterment, of course."

President Ericson waited until a woman in the center of the room was handed a mike. The lady was highly incensed.

"Individuals have little meaning to bureaucrats. When you have a complaint they file your name in a waste basket."

Jodi spoke slowly, thinking back on conversations with her father.

"Individuals with names have built our world as we know it," she said. "Think back," she addressed the audience with extended arms. "Do we remember countries or nations or ethnic tribes? No. Our civilization has been built on individuals with names aptly suited to them. Could the Christ Child have had any other name but Jesus? An odd name like Napoleon seems right for him. Or George Washington, Abe Lincoln, even Genghis Khan. Alexander the Great, Caesar or Columbus. Then there's evil personified in Stalin, Hitler and Tojo. The names seem to fit. It's because individuals create the names which live in history. Governments don't create lasting memories--individuals do.

"So my friends," said Jodi. "Individuals such as yourselves do count. And the world will remember what you as individuals do far longer than it will remember this congress."

A lady in a flowered dress in the front row had a question.

"Madame President," she asked in a quavering voice. "What can single individuals do?"

"A worthy question," said Jodi, walking to the edge of the platform and speaking directly to the woman. "Get together with neighbors. Organize. Demand your congressman and senator pay attention. Insist they lighten your tax load. Let them know from now on your load is also their load. No more excuses or diversions. If they don't respond, get another horse."

Jodi smiled. "As a young girl riding the state fairs I learned the best way to win a race was lighten the load on a horse. Handicap races do the opposite. They load the best horse with the heaviest load until it's brought down to the level of the worst horse. Every teenager knew that if you had a good horse, the extra weight lowered your chances of winning.

"Now congress has adopted that philosophy as a national policy...load industry with taxes and regulations until they collapse in an unfair struggle against weightless international competition.

"If a teenager can understand something so simple, why can't congress? It's up to us individuals to awaken congress or get rid of them entirely. Follow the example of our forefathers with the inept Continental Congress.

"We need new racing rules. If we must discard congress to eliminate dead weight, perhaps we'll be better off. That's why I seek your support for my XXVIII Constitutional Amendment."

A freckled face young man in the back hollered his complaint.

"Congress is bleeding us to death with hundreds of minor cuts. How can we stop them?"

"You're right," responded Jodi. "Hidden increments are sapping democracy's energy."

"Meaning what?"

"Withholding taxes are hustled out of wages and salaries in small increments. Health care costs, unemployment insurance, income taxes, estimated taxes and fees are extracted in bits and pieces. Car and housing taxes are paid in monthly increments."

"So what?" The freckled youth was angry.

"If people had to pay all taxes April 15 instead of by increments they'd realize how deep their wounds and how much total blood they're losing. The shock trauma would create a march on Washington which would force congressmen to cut taxes or flee for their lives. The French Revolution would be comparative child's play."

"Where'd people get the money to pay huge lump sums on April 15?"

"They'd have to borrow it. From banks and other institutions."

Jodi raised her hand to quiet the protests.

"If voters had to pay interest to banks to borrow in order to pay taxes, they'd launch a tax-reduction typhoon across the country which would sweep away perks, privileges and special interest boondoggles for another 100 years."

"We sure as hell need a typhoon," hollered a man in a baseball cap.

"Right now we have a deficit of $200 billion and rising," said Jodi. "Fast. Benjamin Franklin investing $100 would be a billionaire today. That's the kind of compounded debt we're leaving our children and grandchildren. They can't possibly escape such a future tax load. They will be economic slaves no better than the victims paraded in chains through the streets of Rome by conquering Emperors."

"Your ideas would create another civil war," cried the questioner, still on his feet.

"Our civil war hammered us into one nation," said Jodi. "To keep this one nation requires continued courage and vision. Taxes, ethnic, parochial or racial self-interest problems are devisive enough to send us back into another destructive civil war."

Jodi grabbed the microphone and lifted it away from the podium as she strode across the stage.

"Like the old AESOP fable. One stick can easily be broken; not so with bundle of sticks. If we allow ourselves to be separated into ethnic and gerrymandered sticks, we'll surely be broken as a nation."

Jodi stood quietly before the confused assembly before issuing her tongue-in-cheek solution.

"Taxpayers can turn April 15 into another disastrous French Revolution. Or they can wake up now and prevent one."

43. NOVEMBER 20: INTRIGUE

Speaker Metzer, Senator Meyer and Reverend Rowan spent the entire evening over dinner at the PUB restaurant one block from the Capitol. The restaurant was one of the city's finest. Famous for large portions--only the circus fat lady left the place hungry--the PUB was a favorite watering hole for the media and members of congress.

The Chairman of the House Judicial Committee pounded the table with joy. The Speaker cautioned Jesse to keep his voice down as diners raised their eyebrows at his excited vehemence.

"The black caucus, with a single exception, is solidly behind us on impeachment," declared Rowan. "Also got 132 calls from members who had the rug jerked out from under them by the President's budget veto."

Senator Meyer chuckled.

"Seventy percent aren't back in town yet. Wait until TV takes the impeachment circus public."

"That's certain to steal the spotlight off the President's desperate tour to sell us down the river," grinned the Speaker. "Once the Nixon show got on television no one would have noticed had Hitler risen from the dead. People were glued to the impeachment coverage on TV."

"That bitch insulted the black caucus," exploded Rowan.

Mordecai glanced about nervously.

"Take it easy, Jesse. Let's not turn this into a racial war. How'd she insult you?"

"Vetoed our priorities in the budget, impounded additional funds for food stamps, welfare, housing subsidies for inner-city poor, and financial payments to pregnant high school and college kids."

Senator Meyer, convinced Reverend Rowan wanted to keep alive government programs more than he wanted to solve inner city problems, said sourly.

"None of which is paid for by your constituents."

"Doesn't matter," said the Reverend. "Pass the collection plate only after you've pumped people emotionally. Reach deeper into their pockets that way."

Senator Meyer glanced up to see Senator May Trudeau entering on the arm of General Pete Hardin. He was dressed in a blazer blue jacket with white shirt and an Air Force blue tie. Senator Trudeau was stunning in a solid blue jersey dress which clung to her figure like a wet sheet. A golden rope belt snuggled her petite waist and accentuated her full bust.

Mordecai had an idea.

"Perhaps they'll join us," he said, jumping to his feet as they approached. He stepped in front of May as she followed the Maitre d' to a table.

"Senator Trudeau. Could we entice you both to have a quick drink with fellow congressmen? We'd be delighted."

The suggestion caught Senator Trudeau by surprise. By this time Speaker Metzer and Congressman Rowan were also on their feet pulling out vacant chairs.

Momentarily disconcerted at an invitation from a relentless adversary, Senator Trudeau turned questioningly to General Hardin. He smiled. Many times he had negotiated cordially with an enemy before battle.

"Why not," he said. "But only for one drink. The Senator and I have business to discuss."

Monkey business, thought the Speaker. But he said.

"We'll not detain you long. We've finished eating."

Senator Trudeau slid into her seat slowly, not taking her eyes off the Speaker. She knew he was an implacable foe of her friend, President Jodi Ericson. Could there be an underlying motive?

The awkwardness of the situation had no effect on General Hardin.

"A first step towards 'peace in our time'?" Pete Hardin questioned cheerfully.

Senator Mordecai Meyer laughed easily.

"Arafat and Rabin sat down together, why not members of congress?"

"Touché," said Hardin. "We've many things in common. I noticed none of you fought reopening military bases in your bailiwicks."

"No," said Mordecai. "Fought against closing them in the first place. Nearby towns and cities love the revenue. Only problem is..." his voice trailed off as he sought to clarify his concerns.

"Yes?"

"The idea of turning them into Youth Work Camps is insane."

"Schools plus camps," corrected General Hardin. "We expect each youngster to graduate with a high school education. A nation of educated kids instead of hoodlums."

"From military bases?" Speaker Metzer questioned with a cynical smile. "Most are far too huge and complex. Maintenance costs per occupant will zoom into the stratosphere. Taxpayer howls will reach the moon."

"Most bases are near colleges and universities. Some quite small. We've contacted many," said the General. "They're delighted with the idea of providing curriculums and teaching certain classes."

"And particularly delighted with the extra income," added Senator Trudeau.

"That's not even the end game," said Hardin.

"There's more?"

"Evidently you've forgotten the President's order for Universal Conscription to start by July 4th one year from now. By then I expect to have all necessary barracks and facilities ready."

"My God," exclaimed the Speaker. "The expense?"

"No more than building immense prison complexes, half-way houses, increasing social workers and caring for teenage mothers." General Hardin accepted his ordered scotch and water from the waiter. "Plus the enormous financial load youthful crime has put on police, hospitals and city budgets. Washington, DC, even wanted to call up the National Guard for relief.

"The military for years has been superb as teachers," Senator Trudeau commented. "Without disparaging anyone, I don't recall teaching as a fundamental accomplishment of the police. Many of today's cops are youths who could suffer a bit of teaching themselves."

The Speaker sat stunned. He knew the President's plans but hadn't expected them to be launched without approval by congress. He queried the General with a frown.

"You're putting the military side-by-side with work camps?"

"Why not? Facilities are in place. Middle-class parents sent junior high school boys to places like Leonard Hall Junior Naval Academy. The boys loved it...fought to gain admission."

"Far better place than alleys and slums," said Senator Trudeau, amused at the Speaker's chagrin.

Reverend Jesse Rowan, despite his normal raucous bravado, had been temporarily subdued by the commanding presence of General Hardin and the haunting beauty of Senator May Trudeau. He struggled to be included.

"Blacks been better treated in the military than in any other element of American society," he admitted, turning to the Speaker. "In my congregation parents say their boys love military life. Best we not fight this program of the President's."

"Got to clean out and fumigate any dump before building a house," said Senator Trudeau. "Means first digging out the loose trash in our schools and streets. A problem Jodi's attacking head-on."

"Roger," said General Hardin. "Get down to bedrock on crime, education and indolent youth roaming the inner cities with nothing to do except use drugs and create teenage babies. Millions spent on social programs haven't worked."

"Federal programs are a disorganized disaster," said Senator Trudeau. "I come from ranch country. Teaching a kid to ride a horse by social preaching only gets him tossed on his can when he straddles the real thing. Small wonder he becomes contemptuous of educators who mumble rote in classrooms of chaos."

"You'll have your hands full of tough youths," the Speaker advised the General.

"The military's never had serious problems with idle trouble makers," Hardin said. "No need to build bigger prisons for the vicious 10 percent if we find outlets for the 90 percent youths who have committed no bodily harm to another. Let's insert the 90 percent back into society after providing discipline, a responsible work ethic, and hands-on education in controlled study halls. Sticking a toe into the water doesn't work. Learning to swim involves the whole body. Today's schools barely educate the toes."

"Aren't you taking a chance of drowning the patient--throwing him into a military lake away from his environment?" called out a skeptic.

"Rarely do youth's learning to swim drown. It's when they don't learn that tragic accidents happen," said Jodi. "And the whole body's got to be immersed in any realistic effort to produce results. Exactly what a Youth Work Camp does. Takes over the entire body from dawn to taps. A new healthy body, physically and mentally, emerges."

Half-way houses are breeding swamps for idle minds," said Senator Trudeau. "Abolish them and put the money in the Youth Work Corp."

"You're breaking individual right laws in establishing your camps, General," asserted Senator Metzer. The President's program threatened many of his pet social projects.

"Coercion, not law, is at the foundation of all social life," said General Hardin grimly.

Speaker Metzer sat back in his chair and studied Senator Trudeau at some length. Like the President she had looks and money. Women today were becoming a potent force in politics but their thinking often evaded him.

He suddenly wanted to know.

"Why stick up for the President so passionately when she wants to take your job...and the jobs of all of us in congress?"

"We're here to serve, not provide employment and special privileges for ourselves," said Senator Trudeau, a bit waspish in retaliation to the Speaker's tone.

"If the American people aren't taxed an additional dime for the next ten years, they'll still pay enough taxes to support any prudently administered government. Members think they have a right to reach into another's pockets for money to execute their own fantasies. Half our social programs are disguised bribes to get members reelected."

General Hardin, sensing the animosity creeping into the conversation, abruptly slid back his chair.

"Senator Trudeau and I must go. We've things to discuss. Thanks for the drinks."

Hardin signaled the Maitre d' and shortly the two were seated at a table in the far corner of the room.

Senator Mordecai Meyer sat punching his napkin with a fork.

"Just as well," he grumbled when they were out of earshot. "Bad idea in the first place, inviting a Trojan Horse into camp."

"Let's face it," said the Speaker. "The real reason we've got to hang the President is she's already put the people in a hanging mood...with us the victims."

"Yeah," said a bitter Reverend Rowan. "An old time lynching. I feel the rope already."

A disturbed expression crossed Senator Meyer's face.

“We hang her or she hangs us? That the ticket?”

“I’ve known that bitch for years,” said Metzer, his eyes glinting with hateful memories. “She believes in getting her opponent to the guillotine first.”

“How do we beat her to the punch?” Congressman Rowan felt helpless anger as he glanced at the table where General Hardin and Senator Trudeau were laughing.

“Enlist the media. President Ericson has thoroughly alienated them with her hard-ball rejection of anonymous sources and her outright refusal to answer personal questions.”

“Trying to influence the media is damn treacherous,” cautioned Mordecai. “I’ve stepped into their dens. Once inside, like rattlesnakes, they’ll turn on you.”

“Make them think all ideas are theirs,” said the Speaker. “Thirty years of dealing with the White House Press Corps taught me which Nintendo buttons to punch. The media has a litany of grievances. Work on them. When their own gripes are involved, they care less about particulars. Any malevolent quote will do.”

“Lots of buttons to push,” laughed Congressman Rowan. “Each media hot shot likes to press his own pet button. Otherwise he won’t join our lynch mob.”

“Right. Every hanging needs a perceived wrong. We help them find the wrongs.” The Speaker slapped his friend on the back.

“Jodi Ericson will be hard to bring down,” said Mordecai. “She’s beautiful, rich...”

“Two reasons she’s vulnerable,” said Rowan. “From the pulpit I’ve learned how to feed envy. Most women are neither beautiful nor rich. A beautiful Marie Antionette ‘let them eat cake’ or a ‘rich people don’t pay taxes’ Leona Helmsley evokes absolutely no sympathy. But an Anita Hill draws tears and money with her phony tale of working girl harassment.”

“Well, I’ve started the ball rolling,” said Speaker Metzer. “I’ve sent a force into the field to dog and badger Jodi Ericson at every event. I’ve alerted congressmen in each district that Jodi’s single purpose is to throw ‘*the rascal*’ out of his cushy job. So each member’s got plenty incentive to line up opposition hecklers.

I've told them the worse they make her look the better their own chances of survival."

"We've got to build a cadre of SS troops from fellow members," agreed Reverend Rowan. "Los Angeles, San Francisco and all big cities have thousands of angry voters. We should be able to rouse them behind the cause of impeachment."

"Let's provide the spark," said Metzer. "Inner city tinder is parched from years of dry rot."

"Members love impeachment," said Mordecai. "Takes the public's mind off taxes better than a foreign war."

"Barnum was wrong about a circus," said a gleeful Jesse Rowan. "Impeachment is the greatest show on earth."

44. PRESIDENT VISITS CAMPS

General Hardin's Herculean effort began to fill the Youth Work Camps. He got schooling and training under way. He brazenly drew on military depots for clothing and supplies. President Ericson, operating under the declared National Emergency, gave him full powers to command all necessary resources.

Youths got reasonable haircuts, not close-cropped military. But no earrings, ornamental bracelets, fancy shoes or leather jackets were permitted. All '*cadets*' wore the same type uniform with stripes according to their grade. Younger boys were separated and protected. There was no hazing. Each youth was given a $300 year's credit at the commissary to purchase snacks or personal items.

A fund of $1,000 was established for each full year's attendance to be received only upon graduation.

Weeks before Christmas results began to show. Streets were no longer filled with idle problem youths. Teenage pregnancies fell 70 percent. Street crimes by youths--aided by strict gun law enforcement--dropped 80 percent. General Hardin set up units of Shore Patrol or Military Police who worked with schools in picking up dropouts. Organized protests and harsh complaints fell on the deaf ears of General Hardin until they gradually subsided. And wonder of wonders, the youths, after an initial period of tough adjustment, gloried in new-found pride. They loved the Sunday parades to which parents and friends were invited after the basic training period was completed.

Finally, General Hardin called President Jodi Ericson.

"Madame President," he said, "with Christmas only ten days away how would you like to visit a Youth Camp?"

President Ericson, who had made a zealous effort to stay out of General Hardin's hair while he was going through the throes of establishing the Youth Corps, was delighted.

"May I invite Senator Trudeau and Tony? They've ardently been following your progress."

"Of course. It's your idea. Bring who you wish."

That Sunday, President Ericson walked down an inspection line in Norfolk, Virginia. She stopped opposite a smiling youth with a missing tooth.

"Where you from, soldier?" she questioned the fourteen-year old. Jodi had been advised by General Hardin to use that salutation with all youths. Boosted their macho image in addition to its disciplinary value.

"Washington, ma'am."

"District of Columbia?"

"Yes ma'am."

"Lose your tooth in training?"

The youth's grin widened.

"No ma'am. Lost it in a fight back home. Not allowed to fight down here."

"Training too tough? Would you rather be back home?"

"Training's tough, but back home's dangerous. Rather stay here. Study too much but food's great. Gained fifteen pounds," he added proudly.

Jodi turned to General Hardin who stood alongside with Senator Trudeau and Tony.

"How about sending this young man to a dentist?"

"No facilities, Madame President. Military hospitals are overtaxed and we'd have to draft dentists...a sticky wicket. Got plenty doctors but not dentists."

"I've looked into the problem for General Hardin," said Senator Trudeau. "Drafting dentists or professionals would take two years to get them up to speed. And the hell raised would make you think you were quarterback in a superbowl. Costs would skyrocket. Dentists need hygienists, dental chairs, X-rays and millions in equipment and instruments. None of this is possible before half these kids graduate."

Jodi surveyed the puzzled eager recruit before her.

"We'll get you a new tooth," she promised.

"Thank you. Could sure use it."

After inspecting all units, President Ericson expressed her pleasure at the job done by General Hardin. But she voiced a nagging problem.

"That kid with the missing tooth? How to get dental help?"

"The time frame's practically insurmountable," said Senator Trudeau.

President Ericson walked ahead without comment for a few minutes. Suddenly she stopped and confronted General Hardin.

"Monday, start sending these kids to civilian dentists. Pay for them as for any ordinary patient. Set up a special fund, I'll see you get the money even if I have to cancel a piece of hardware."

"Could end up a lot cheaper than building new clinics all over the country," agreed Senator Trudeau. "And one stealth bomber now runs nearly four billion dollars. Four billion dollars could fix every mouth in the Youth Work Corps and the military. Using private dentists will not cause any economic dislocations. Sounds like you've hit on a great idea."

"Not good enough to sell to the Speaker or Majority Leader, however," said Jodi morosely.

In the mess hall President Ericson sat between cadets from West Virginia and Maryland. The white youth, James Randolph, was from Morgantown; the black youth, Leroy Jackson, from Baltimore.

"You boys learning anything?"

"I'll say," said Randolph, a pint-sized enthusiastic youngster with a deformed left hand. "Teachers are neat. One officer's from West Virginia."

"Least we got books," said Jackson, a tall black youth who had dropped out of school.

"You had no books at home?"

"One to 'bout every three guys. Could hardly call them books. More like old papers marked up and pasted together."

"Pretty sad school system which didn't furnish students with books," said Jodi.

"Don't matter none. Couldn't read them nohow."

"You can't read?"

"Not hardly. Teacher always way ahead so gave up."

"How about here?"

"Doin' okay," said Jackson. "Instructors give me an' others special tutoring till we catch up. Says I'll catch my own class in another six months."

Jodi sat quietly, a bit shaken. She'd never quite realized the struggle required to learn. Or the patience required to teach.

"Learning the army way's neat," said Randolph. "My officer friend from home says, 'no hurry. We do it over and over and over again until you'll remember when you're a hundred.' Then he calls on us to show him. In school never got called on once."

"Me neither," said Jackson. "Teacher only called on the smart ones to stand up and read."

"Nobody gets to hide in our classes," beamed Randolph. "Each takes turns. Instructors patrol aisles like bird-dogs."

"Are the exams fair? As much as in the schools?"

Jackson snorted his disgust.

"Exams never fair in school. Pass you no matter. Teachers don't want to handle dumb kids."

"Perhaps they don't want to see black kids wind up in the bottom?"

"Bunch of bull. Here no top students. Instead of passing us dummies," he stopped to flash an unashamed wide grin, "instructors sit down and work with us 'til we understand what they're talkin about. Spent first six weeks learnin times table. My sergeant in no hurry. Said he didn't get paid by the hour, nohow."

Jackson's face brightened.

"Know tables backward and forward--all way through twelve."

"Great," said Jodi. She laughed. "Have trouble remembering them myself. Specially when shopping."

"School teachers tried to hide us dummies. Sergeant makes each of us stand up and read in turn."

"How about discipline in the classroom?" inquired Jodi, biting into an apple.

"Real bad at first," frowned Randolph. "Military soon took care of that."

"Had trouble myself," grinned Jackson at the memory. "But like basketball, even idiots learn referees don't give damn about excuses. Marching around mess hall on an empty stomach a few times learns you real quick--you're the only dude going without dinner."

"How do white and black boys get along?" Jodi asked.

"Okay," said Randolph, helping himself to his third portion of green beans. "No longer problem."

"No longer? There once was?"

"Gangs hung together for couple weeks. After that got too tough."

"What happened?"

"Separated and assigned us quarters by grade and alphabet. Between studying and drilling got no time to run across compound to rap old buddies. Made friends at home."

"We live in a different world," said Jackson. "Haven't rapped old friends for least three weeks."

"Whites and blacks get along okay?" Jodi feared that would be a major problem.

"Whitey got nothin' I want," said Jackson. He pulled a bundle of commissary chits from his pocket and waved them before Jodi. "Saved more chits than anybody in my barracks," he said proudly.

"No fancy clothes or jewelry in the commissary?"

"Naw. Where'd we show them off? Dudes all in camp too. Anyhow, girls respect uniforms more than dumb trinkets."

"Sounds like you really have entered a different world," smiled Jodi.

"Better one, too," Jackson said, flashing a quick grin. "Dude marching behind got no reason to stick a shiv in my ribs for a pair of shoes that won't fit him nohow and he can get for free."

The news always played up a nagging question Jodi wanted answered.

"Truant kids once filled the streets. What happened?" she asked.

Jackson and Randolph glanced at each other before Jackson shrugged, answering.

"We're both truants. Schools squealed to Youth Work Patrol which picked us up."

"Why? In the District 8,000 students didn't show up for classes?"

Jackson hesitated a long moment before answering.

"Kids don't need the trouble. Attacked if we went to school. Had no hot-shot clothes to wear. Teachers got plenty to do keeping order without paying attention to me."

Randolph said. "Easier to rap with friends than try to stay awake in school."

"Peer pressure?"

"Guess you could call it that," Randolph said sheepishly. He brightened. "But all those dudes here now. Not on streets."

"Yeah," said Jackson. "Here's better. On streets we start running around with older guys who get into burglaries, car thefts, robberies and do dangerous things."

"Got to steal to have stuff to show off to girls," said Randolph. "Got to go along. Else they think you're no-count creep."

"Guys want to do what other guys are doin'. Here we're all doin' same thing," said Jackson. "So no problems. Even easier. No cop worries."

After lunch President Ericson took a detailed tour of the camp and compared notes.

"You've done a magnificent job in a short time," President Ericson told General Hardin.

"Kids have done the job themselves," said the General, his blue eyes twinkling as he glanced across at Senator Trudeau. "Most kids are diamonds in the rough. Some require more polish than others."

"Abrasive?" grinned Tony.

"Polishing is an abrasive operation. Especially for diamonds in the rough. A form of necessary coercion."

"Shades of Justice Oliver Wendell Holmes," smiled Senator Trudeau. "He said coercion lies at the foundation of all social life."

Jodi thought about the Speaker and Majority Leader.

"General," she said. "The plan is working like a Rolex watch inside the camps, but the people don't see these kids. Weekend visits by friends and parents won't offset the bleating of Meyer and Metzer."

"Don't underestimate our people. They've already noted the difference in their streets and neighborhoods. Hartford, Connecticut, cut crime in half by a 10 o'clock curfew. The people loved it, despite loud dissenters--those who complain about everything and solve nothing."

"Right," exclaimed Senator Trudeau. "In slightly over three months you've grabbed crime at the womb. No one's ever tried going at the source before. I think it's working."

"Kids are quick learners," said General Hardin. "Adults seem always to miss just how quick."

Jodi had one item tugging at her like a nagging headache. It was the military. Universal conscription. Bringing into all these bases young high school graduates for three years' military service. Piled on bases already supporting the Youth Work Corps, how will the mesh work? Surely none would know better than General Hardin.

President Ericson raised the question.

"Pete, when you load all the additional youth into the armed services, will you face an insurmountable problem?"

"No way," he laughed. "Induction centers have faced identical situations numerous times in our history. The Youth Corps really isn't that large. And it's becoming smaller as kids in school are opting to behave themselves and stay put."

"You know I've canceled the Labor Department training money?"

"Excellent idea." General Hardin laughed. "They train sled-dog drivers for jobs in New Mexico."

Jodi questioned. "What about the military itself? Will they resist?"

General Hardin's response was one to which he had given considerable thought.

"Best thing that ever happened. Assured careers. Until now finding enough new qualified people to accept a military career has been difficult. The cutbacks, downsizing and wholesale destruction of military careers previously made quality recruiting damn near impossible."

"Kids no longer trusted the military to give them a lifetime commitment?"

General Hardin nodded.

"Many were forced out despite recruiting promises. Labor Department training was blind to career needs."

"That department was never oriented to more than an aspirin solution, anyhow," said a disgusted President.

"Computers, technical communications, sophisticated weapons force new recruits to be able to perform a variety of duties with high grade skills. Today we need the best and brightest. To get them we've got to be capable of educating them ourselves and keeping a career commitment to them once they're under our care."

"You think Youth Camps and Universal Conscription can do that?"

General Hardin was positive.

"Programs provide a broad pool to get our youths acquainted with a central pillar of our constitution--*'provide for the nation's defense.'*"

"In recent years young Americans have become less interested in joining the armed forces, that's for sure," said Jodi.

"The services give youths such a wealth and breath of training, I still can't fathom why they get such a bad rap?"

"Ignorance," said Jodi. "Ignorance caused by unfamiliarity with the product. Just as we need one language and one currency in the nation's malls, so we need one defense establishment which crosses all lines."

"Well, President Ericson, I'm certain you've taken the right steps. Success will follow," said an upbeat General Hardin.

"If impeachment doesn't throw me out of office first," said Jodi, biting the corner of her lower lip, "we've got a winner. It's a horse race against time."

45. PAST IMPEACHMENTS AS MODEL

The group sat in a corner of the PUB restaurant and plotted the next move on the impeachment of President Jodi Ericson. The five included Speaker 'Fat Albert' Metzer, Congressman Jesse Rowan, Senator Mordecai Meyer, Secretary of Labor Harvey Hyman, and Howard Bloomberg, fired EPA Administrator.

"Best we review past jobs to see which routes were the most successful," said Reverend Rowan. "No use repeating mistakes."

"Sound idea," said Senator Meyer with a slight dig at his none too favorite crony from the House. "Since you're Chairman of the Judiciary Committee I'd expect you'd have a slew of researchers already on the job."

"Way ahead of you," said Rowan. "My Rhodes Scholar buddy knows all about research. He's been giving me a helping hand."

"Purely investigating past successes and failures," said Harvey Hyman, spreading his hands in a self-deprecating manner. Although a member of the President's Cabinet, the Labor Secretary was in direct conflict with the President's policies but hid his disapproval. Instead he worked with Bloomberg to slip inside information to the Speaker, hoping to advance his opportunity for a Senate seat after Jodi's impeachment.

Hyman adjusted heavy black bifocals which he parked on the end of his nose. A bearded face with shaved neck gave him the look of a plucked chicken.

"Research indicates we're winners. Hands down."

"Why so cocky?" Senator Meyer thought the effete scholar a dilettante who aspired high places without getting his hands dirty.

"Past impeachments had little to do with merit, only emotions," said Hyman. "Congress is stomping at the starting gate to impeach our arrogant cover girl. She's proved William II's remark 'she's so beautiful that to expect sense from her would be hoggish.'"

Reverend Rowan wiped his nose on the back of his hand.

"Skip the fancy talk. An old black proverb says it best. 'White hen mighty pretty, but de black chicken is de luckiest when de hawk come sailin' ovah de chicken yard,'" he chuckled. "An' de hawks in congress is swoopin."

Senator Meyer smiled despite himself.

"Good point. Still got to come up with solid charges."

"The three you advanced are plenty," said Hyman. "Now we'd best concentrate on beating her with emotional punches."

"Using emotional charges let Andy Johnson get away," said an unconvinced Senator Meyer. "Historians claim the Johnson impeachment 'was one of the most disgraceful episodes in our history. No valid ground, legal or otherwise, existed for impeachment. It was a great act of ill-directed passion.'"

The Majority Leader realized there had to be substance or at the last minute seasoned senators would have second thoughts.

"Impeachment of Andrew Johnson was mainly an effort to change the policy of the government by removal of the executive," said Hyman. "Johnson wanted to carry out Lincoln's forgiveness. Northern radicals wanted Southerners to pay for their sins. Since Southerners were excluded from congress, it was no contest."

"Reason we've got to have substance," said Mordecai. "Today those Southerners are back on board. Always expect a hidden few moralists to surface on principle. Then all our work's gone for naught."

"Ultimately the question was not so much whether President Johnson should be removed *because* of committed offenses, but whether he should be removed for the public good," said Hyman. "That's really our goal today."

"Justice Brandies said nothing is clearer than the intention of the Founders to repudiate and reject 'legislative omnipotence,'" worried Senator Meyer. "After the civil war congress executed a series of usurpations intended to make the majority the ultimate judge of its own powers, and the President merely responsible to congress. When that objective became obvious, the Senate balked. Let's not let this impeachment become another emotional fight. Emotional warriors always find themselves in a battle without weapons."

"Votes are votes, no matter the motives," said Hyman. "Charles Sumner fanatically hated Johnson, yet he sat in judgment."

Said Mordecai. "We've got to be careful of a purely emotional impeachment. Some senators won't stand for it."

"Impeachment of President Ericson is not the problem," said the Speaker. "I can handle that. Only a simple majority is required in the House for impeachment. But when we go to the Senate for conviction we've got to muster a two-thirds vote."

Mordecai Meyer worried. "The Senate's loaded with independent mavericks. I don't have your sheep to control."

Speaker Metzer caught the Majority Leader's jibe.

"But you've got a smaller herd to handle. You can deal personally with strays."

"Violations that justify removal from office demand proof," insisted Senator Meyer.

"So how do we hang Jodi Ericson before she hangs us?" The Speaker was determined to pursue impeachment at all costs. He was convinced the battle had gone beyond any compromise with the President. Her articles of amendment made that abundantly clear.

Secretary Hyman had the answer.

"Who wins the battle of emotions wins the war. Gin up the media. Andy Johnson came only one vote shy of conviction long before TV or radio talk shows. If we can get the media to jump on Jodi, she'll sink quicker than the Titanic."

"The media loves picking on a wounded chicken," chortled the Speaker.

"Still need substance," said a skeptical Senator Meyer.

"We've got that too," said Hyman. "Far more than Nixon. His came down to a single charge of obstructing justice in connection with the FBI investigation of the Watergate burglary."

"Only proved by the tapes," said Mordecai. "We've got no tapes."

"Don't need any," said a smug Harvey Hyman. "Jodi Ericson has committed far more transgressions than Nixon."

"I'll say," echoed the Speaker. "She stands convicted by her own mouth."

"This President's flaunted more laws than any since Abe Lincoln," said Hyman. "At least he had the excuse of a civil war."

"Her basic depravity is her contempt of congress," said Metzer bitterly. "She holds us in lower esteem than a snake's belly."

The Speaker's hate was intensified by his uneasiness over his handling of the rape incident years ago. Until Jodi was vanquished from the scene completely, his mind would never rest. Like Javert in pursuit of Jean Valjean in Les Miserables, Speaker Metzer's personal failure had only intensified his hostility.

"Let me count the sins," said Hyman. "In addition to her oft declared contempt for congress, she's invoked a state of National Emergency; insultingly chopped off funds voted by congress for such things as prisons; created a Youth Work Corps to be run by the military without any law establishing such an operation; strangled and closed agencies with budget restrictions; called up the military to surround the Capitol and kept congressmen from entering their own offices; initiated an amendment which would eviscerate the powers of congress and repudiate the checks and balances of the constitution."

"Worse, she intends to dismantle the entire Welfare State," savagely injected Reverend Rowan, who had been silently listening. He had no knowledge of U.S. history and cared little about it, although he was sensitive to the fact the black race was the only race without a written history.

"A damn formidable array of charges," said the Speaker. "The Senate either convicts or gets tossed out on its ass like the Continental Congress. In thirty years, I've met few senators who wouldn't commit mayhem to keep their jobs."

"Exactly my point," said Hyman. "Our arrogant young Joan of Arc faces certain conviction. Her judges harbor pure animosity."

A broad smile crossed the Speaker's face.

"Can't imagine any greater joy than seeing that young snip swinging from the yardarm of impeachment. What a great night's sleep I'd get."

* * *

In Washington the next evening, President Ericson had just finished a light meal with Tony when she received a call from General Hardin.

"Senator Trudeau and I are just leaving the PUB. How about a quick stroll down a few streets in Washington? See what you've accomplished?"

Jodi hesitated a moment. She glanced at her wrist watch. Nearly nine. Normally not the time to walk the city's streets.

"At this hour?"

"None better. That's when the action is." Jodi heard the General's chuckle over the phone.. She knew he was tossing her a dare.

"Why not? I'll bring Tony in addition to the Secret Service."

"We'll pick you up in my jalopy. That way, we won't disturb the natives."

"Two hours later the foursome had toured 14th Street, scene of the riots; an area east of the Capitol; several blocks near the river; and the poorest areas in the city.

"Nothing much happening," said Jodi. "Beautiful night.

You'd think kids would be out in droves."

"My point exactly. District streets are calm and quiet. They weren't this tranquil under the National Guard. Months ago this area teemed with gangs and drug pushers. Now you can throw a baseball in any direction and not hit a single juvenile. "

"My god, that's true," exclaimed Jodi. "I hadn't noticed. My mind was on tomorrow's tight schedule."

"The people out enjoying the night air are all older," said May. "What a relief."

Two policemen on the beat waved from across the street.

"Patrolmen are on their feet greeting the people," said the General. "Squad cars cost three times as much and alienate the neighborhoods. In another year they'll probably be picnicking together."

Jodi stopped and took her time surveying the situation, a broad smile spreading across her face.

"Good lord, General. I think you've accomplished the impossible," said Jodi. "No one in his right mind would have dreamed of taking this walk a year ago."

After visiting a few more serene neighborhoods, an invigorated President Ericson returned to the White House.

"General, you've given me one hell of a shot of adrenaline. Once the country wakes up to your accomplishments, impeachment will only be a dirty word."

Unfortunately, neither the Speaker nor Majority Leader had accompanied the strolling party.

46. CHRISTMAS REPORT

President Jodi Ericson decided to make a final report to the people before permitting congress to reconvene in January. She had changed her mind and relented since adjournment in August. The National Emergency had enabled her to place in the works her major programs and nothing much except bad publicity could be gained by excluding members from their seats. She dismissed the National Guard, substituting a skeleton military force on call at nearby Bolling Air Force Base. Jodi again permitted the Capitol Police to take over. Occupying offices gave members no more power than they possessed meeting in Washington hotels.

"Give them the capitol forum," she told Tony. "I'll continue to use TV and radio in direct talks to the people."

"Fanatics like Reverend Rowan end up turning people off," said Tony. "Better to let them drown in their own offal than contaminate us."

Jodi smiled at Tony's vehemence. He had little regard for the whining supplications of Congressman Rowan for special treatment.

Tony commented. "Reverend Rowan's like a fighter bellowing how he's gonna knock out his opponent in the first round. Then in the ring he pleads with the referee to make the other fellow fight fair, quit hitting at the breaks, tone down his savage attacks."

"Ultimately, it's up to the American people to carry their own fight," said Jodi. "If this democracy's to survive, they've got to take charge of their own destiny. You and I can only hope to provide an honest playing field. One without pot holes or hidden weapons."

"We called them 'nukes' in the fight game," said Tony with a tight laugh. "Steel knuckles wrapped under tapes."

Jodi nodded. "Special interests hidden from public view?"

"A deadly advantage."

"An advantage we've got to take away from congress. They've been using hidden 'nukes' on the people far too often."

"No wonder the taxpayer's groggy. He's been punched around with hidden taxes and camouflaged catastrophes so long he can't tell the needy from the greedy."

Jodi smiled. "Plan to change that with my *Report To The People* Sunday. Marvin Pringle's promised I can deliver my message without guests or interruption."

"You're taking one hell of a chance," worried Tony. "Congress, not the people, is the one to worry about. They got the power to impeach you."

"If the people let them, so be it," said Jodi with nervous laugh. She knew the odds were high, the stakes even higher. "If our democracy's to last another 200 years now's the time to find out."

Tony shook his head dubiously, his eyes troubled. His love for this one-time waif had mounted to a Joan-of-Arc passion. Money had not changed her. Nor had the power of Presidency. Despite adversity, Jodi had maintained a Mongol's courage and reckless spirit against sometimes frightening odds.

"For their own sake, I hope the people respond," said Tony.

"I'm firing the first salvo. They can get behind me or slide into political slavery. A people that allows itself to be over-taxed, over-regulated, disarmed and misruled by Washington needs to face up to its very existence. Can we really rule ourselves? I'm not sure."

*   *   *

Opening his Sunday TV program, Marvin Pringle made a simple introduction.

"President Jodi Ericson wanted to make a Christmas *REPORT TO THE PEOPLE.* I'm delighted she chose my program. May I present the President of the United States."

Jodi stepped up to the podium from behind light blue studio curtains. Smiling her acknowledgment to Marvin Pringle, she carefully arranged her notes on the stand before looking out into the studio. Jodi preferred live audiences because they prevented a speaker from dropping into a mechanical monotone of script reading.

"Fellow Americans," she began, slowly enunciating each syllable in the manner of midwesterners. "Like you, I've been reading in the papers what an unfeeling dilettante you have for President. I've insulted members of congress by suggesting they help their employer--you the people--survive. By helping you stay afloat by throwing out a life jacket, not chains."

Jodi chuckled in her low tomboy voice.

"Their idea of a life jacket for you is a luxury liner for themselves.

"Citizens who devote daily efforts to eking out a living have little time for national problems. That's why you send representatives to Washington to devote 100 percent of their time to solutions. But always their answer's been to create more programs requiring more taxes and inventing more problems. Why do they do this to us?"

President Ericson let the question hang in the air. She took a sip of water.

"Because special interests take over and side-step your interests," she said. "Despite all the veiled excuses and yap-yap the problems are clear enough. In case members profess ignorance, let me spell them out. Let these be known as the People's Ten Commandments demanding action. They should be pasted on every congressman's door. And every voter should know them by heart and judge his or her representative accordingly.

"1. Spending--reduce.

"2. Taxes--reduce.

"3. Crime--reduce.

"4. Education--improve.

"5. Immigration--control.

"6. Bureaucracy--reduce.

"7. Foreign Policy--enunciate.

"8. Defense and Domestic Security--provide.

"9. Lawyers Control of Legal System--abolish.

"10. Quality in American Life--improve.

"Each ***Commandment*** needs addressing in a different way, which my constitutional amendment seeks to do," Jodi continued. "But your job is to hold representatives' feet to the fire with your votes. Ask direct questions and insist on specific answers, not arm-waving generalities and promises.

"In my talks across the country I plan to explain my views on each of the Ten Commandments and hope you will agree. Meanwhile I'm going ahead with solutions under authority of a National Emergency. This nation needs solutions, not more congressional finger-pointing hot air."

Jodi Ericson gripped both edges of the podium and took a deep breath. She straightened until her dark hair fell across her shoulders behind her neck. Her erect figure reminded onlookers again what an extraordinarily beautiful woman they had as President.

"Now to a personal matter which you've been bombarded with by the media--the matter of my impeachment.

"Impeachment by vindictive partisans is no better than a Soviet-style Kangaroo Court under Stalin. Nowhere in the proceedings is the accused allowed to defend himself. So I'm defending myself before the American people. Taking my cause to the airways. TV, radio and public forums will convey my message. Democracy is the people's government, so let the people decide. Send your votes to members of congress. Let them know a vote to impeach is a vote to return home, where they should have spent more time in past years.

"Tell your representatives to move back home where they belong. Living for generations inside the Washington beltway has contaminated their thinking; removed them from local schools, regulations, taxes and problems in the districts they're supposed to serve. Shorter sessions in fancy Washington restaurants and more time with local tradesmen might give members a fresh outlook."

Jodi smiled at the studio audience, which was listening with rapt attention. She hoped the nation also heard.

"Let me assure you, fellow Americans, impeachment proceedings are another giant Houdini camouflage. At which your elected representatives are becoming quite adept. Like Napoleon, Hitler or Mussolini, attack anyone--the Jews or helpless Ethiopia. Divert attention from dealing with tough internal issues."

Jodi looked at the friendly faces of Tony and May smiling encouragement. Tony signaled 'right on' with circled thumb and forefinger.

"A government has three major functions," said Jodi, refreshed at the sight of her champions.

"Defend the country from external attack, maintain a sound currency, and assure domestic order. Tell me. How has your congress responded?

"To defend the country, it has reduced the defense budget one-half."

"To produce a sound currency congress is enroute to destroying a capitalistic system of free enterprise and introducing a spend-and-tax policy which ultimately will drive the nation into a slave state. Even a tyrannical Soviet Union couldn't forever spend beyond its income. But congress persists in following Pied Pipers with megaphones because they're convinced the people are too stupid to handle their own money.

"Assure domestic order? Now that's a real laugh," said an angry President.

"Private security is one of the fastest growing industries in the United States," Jodi declared. "Consciously or unconsciously congress feeds crime in this country by excessive pandering to lawyers. Nobody in history has done so much damage to U.S. business. Or put so many criminals back on the streets as lawyers.

"Ordinary people know democracy does not work in their favor when lawyers enter the picture. Everywhere lawyers have disrupted justice, convoluted the interpretation of laws, engineered prolonged costly settlements, interfered with administration, and made it impossible for middle class citizens to obtain simple justice from our courts.

"By the time congressional staff lawyers finish with a piece of legislation, to be further dissected by lawyers in conference committee, the legal briefs are pages long. In most cases a single paragraph worded in plain simple English would suffice. Plain English worked for our Constitution. Plain English would work again if we threw out obtuse lawyer legalese.

"Today legislation requires entire law firms to reinterpret, at $400 per person per hour, the meaning of an obscure phrase written in pompous legalese by a staffer barely out of law school.

"One columnist said. 'A Thug in Prison Cannot Shoot Your Sister.'

"I say. No. But a thug sprung from prison can. Under our system all thugs get out sooner or later. Even foreign countries want nothing to do with our lawyers. So far they're smarter than we are. Russian people today have called U.S. lawyers *exploiters* and want them sent home. Religious leaders declare that as the power of lawyers grows the moral strength of religion shrinks.

"Congress turns my stomach with their incessant pleas for money. No crime wave driven by hard times emerged during the Great Depression. Poverty then ravaged people ten times worse than this decade has ever experienced.

"Which is why I offer a different solution."

Jodi took a sip of water while she surveyed her audience. They were still attentive, no sign of restlessness. She studied her notes.

"Youth Work Camps run by the military. We should build, not deplete, our military manpower base. The Labor Department wants to retrain youths: to do what? Fry hamburgers? Post security guards in shopping malls? Costs of Youth Camps would not equal one battleship, one Stealth Bomber, or one aircraft carrier. Proposed new prisons are superfluous if we confine the use of present ones to hardened criminals.

"The military's success in building black advancement far exceeds that of any social program. Arthur Ashe remarked, shortly before his death, 'military service has been the most successful career training program we've ever known. Don't dismantle it, rechannel it.' And that's exactly what I intend to do."

Jodi brushed a lock of hair from a perspiring forehead.

"General Powell came out of the military. The nation might never have heard of him had it not been for the military. No other pursuit offers such equality of opportunity.

"As in sports," Jodi said, "life in the military demands discipline and teamwork. The military teaches kids in Youth Work Camps who are taken from inner city schools. Best of all, youths love it. They build units of mutual respect, pride and support. Sending kids to prisons is waging another useless war.

"I plan to cut the enormous costs of social workers. This money can better support thousands of youths working. Under no circumstances do we need to raise taxes to employ our youth. But we certainly need to reign in eager-beaver bureaucrats charging into our lives like Hitler's self-righteous SS misfits."

Jodi's face broke into a wide smile.

"Read my lips, no new taxes." She laughed. "Remember that hollow promise. It defeated a President. But my lips are connected to a backbone of steel. Many opponents call me an insensitive unyielding Mongol. They've yet to learn how damn unyielding on taxes a Mongol can be."

When Jodi finished her remarks Marvin Pringle stepped in and signed off for commercial breaks. During the interval he whispered information to Jodi of a phone call just received from Speaker Metzer. The Speaker and Chairman of the House Judiciary Committee wished to challenge the President to an open debate. Speaker Metzer argued the public wasn't getting an honest picture of President Ericson's offenses.

"If they want an open debate," said an incensed President Jodi Ericson, "let's give them one. Up to you, Marvin. It's your show."

Pringle, taken aback by Jodi's vehemence, spoke cautiously.

"You'd face the two of them?"

"Barking dogs rarely bite. It's the slinking coyote hiding in the tall grass I worry about."

"It's okay to announce?"

To Marvin Pringle his program came first. Such a debate would capture the nation's attention and send ratings through the roof. He wanted to solidify the deal before any party had second thoughts.

"Fine with me," said Jodi.

After the break Marvin Pringle, still acting as monitor, notified listeners of the Speaker's challenge and the President's acceptance. He set the time of the debate for the following Sunday, giving each side one week to prepare.

"Never before in our history," said Pringle to the national television audience, "has a President faced accusers in open debate on a matter of impeachment."

47. THE DEBATE

Tony fumed.

"Pringle, your friend, set you up," he stormed, pacing across Jodi's suite in Jersey City's swank Plaza Hotel. She was in the dressing room leading from the bathroom. "Anything to improve the show's ratings. No lamb's exempt from slaughter if it will help the media."

"You think me a lamb?" Jodi chuckled in her soft voice as she stepped around the corner in a huge white towel. "Few moments ago you called me a tiger. Really, Tony, I wish you'd make up your mind."

Tony blushed. "In bed I'm orbited into another world. Takes a good ten minutes on my feet for my mind to function again."

Jodi approached Tony who was prancing around in shorts. The towel was tucked tightly around her breasts, which provided its only means of support. Her dark hair, still damp from the shower, fell straight down her back. Tony stopped short. Her mouth seemed even more sensuous without lipstick, the lips full and moist. His breath quickened as he warily tried to divine her intentions. Without a word Jodi reached around the back of Tony's neck with both hands and pulled his mouth down onto hers.

Her body tight against his, she said.

"Drop my towel, Tony. I think you need another tiger lesson."

* * *

After breakfast, on the way to the Convention Center, Jodi tried a cheerful note.

"Don't fret so much, Tony. They're the ones who've pulled the boo-boo in this debate. Never before in history has the President been able to confront his accusers. They reached judgments in their own sacrosanct chambers. In Nixon's case the House hauled his faults across hundreds of TV hours without him being present to rebut a single word. Guilty or innocent, it's a star chamber operation. Now I'm confronting the ring leaders."

"But changing the site to Metzer's home bailiwick is tossing a Christian into a Roman arena filled with lions."

"Pringle caved in to pressure upstairs," said Jodi. "Speaker's power at work again. License renewal time and Pringle's owners run scared."

"Outrageous cowardice. Nothing unusual for a government agency."

"Governments run more on fear than common sense," said Jodi. "If it isn't voter fear it's regulatory fear."

Tony worried. "You can bet Fat Albert's home television station's got all guns primed."

"Don't fret. Television's a nasty medium...worse than a firing squad. Men have lived and died by it. A lie detector makes more mistakes than television in exposing scoundrels."

"The Speaker and Majority Leader are no strangers to television. They've spent half their political life dealing with the media."

"Like Casey Stengel," said Jodi in a supreme effort to bolster confidence. "The game's not over, till it's over. We play to win."

* * *

Senator May Trudeau and General Pete Hardin were waiting at the curb for President Ericson's motorcade. Both had warned Jodi against speaking in New Jersey.

"People tried to warn Kennedy against speaking in Dallas," Tony fretted. "He didn't listen."

"Can't be that bad," Jodi joked. "Any state which elected a woman governor's got to have considerable brain power." She referred to the election of Christie Whitman.

Jersey City, once the domain of powerful Boss Hague, was now Mecca to Shiite Muslims and backbone of Speaker Metzer's support. Goon squads picketing Jodi's appearance were held back by police lines. Jodi got a fleeting glance of Ali Rahman directing operations from the rear. She stepped out of the car and stood momentarily on the sidewalk surveying the mixed cheering and booing crowd.

Suddenly a youth darted under the arms of a policeman and lunged at President Ericson with a knife.

Brandishing a sharp pointed stiletto he slit her silk shirt open at the neck and down to the waist, popping out both breasts and nicking one which oozed blood.

Jodi's quick thinking prevented total disaster. At the last moment she saw the attack coming and managed a quick step backward.

Her next reaction was instinctive.

Reaching out, Jodi grabbed the bearded assailant's wrist and yanked him violently. Her lurch backward had thrown him off balance. Since the knife thrust carried him forward, it didn't take much strength to jerk him further off balance and face down onto the sidewalk. As he crashed Jodi twisted his wrist inward. He fell hard against the point of the knife, driving it deep into his stomach.

The assailant's howls of agony split the air. Tony rushed up, rolled the youth over and pulled out the knife as blood spurted over the sidewalk and onto his shoes.

"Call an ambulance," he said calmly to the agitated Secret Service. Tony then pulled up the assailant's shirttail and pressed it into the wound to staunch the flow of blood.

Jodi looked up in time to see Ali disappear around the corner of the auditorium. She secured her blouse with a safety pin offered by Senator Trudeau after covering the seeping blood from her breast with a handkerchief.

"I've lost more blood from paper cuts," she said. "Just hope the safety pin holds. Hate to give photographers a field day." She insisted the cut was minor compared to gashes received on the rodeo circuit.

After watching the police shove the attacker into a van, Jodi permitted a paramedic to sterilize the wound and place a first aid bandage over the cut. Jodi marched into the building as if nothing had happened.

"Place is a keg of dynamite waiting to explode," muttered an enraged Tony.

Senator Trudeau informed Jodi of more bad news.

"Pringle's added Reverend Rowan to the debating roster."

"Three against one?" exclaimed Tony. "Tell them to go to hell. Either that loud mouth withdraws or you do."

"Marvin was damn unhappy," Senator Trudeau shrugged. "But Rowan insisted he had that right as head of the Judiciary Committee handling the impeachment."

Tony stormed. "Tell him to...."

Jodi laid a constraining hand on his arm.

"Too late now, Tony. Anyhow, that triumvirate forms the head of the snake. No better chance than to sever them all from the main body in one swipe."

General Hardin smiled. "Daring maneuver. Quick way to win or lose battles."

One thing Marvin Pringle had insisted on. All four debaters were in the studio standing before separate podiums. The Speaker had wanted to conduct customary guerrilla tactics from his New Jersey office.

The station had widely publicized the event and the auditorium was packed. Extra speakers and TV sets were provided in nearby rooms.

Speaker Metzer, drawn by lot, was selected to begin the debate.

He wasted no time. He'd learned early that TV was a demanding master pitting opponents not only against each other but against a merciless clock. Fire your full arsenal of weapons early, no matter how poisonous or irresponsible, and don't worry about justification. By then the program is usually over. Memories remain with the charges, not the answers.

"Never in the history of the nation," said Metzer, "has a President so arrogantly and flagrantly committed acts with such disdain for our laws. Her declaration of National Emergency; preventing duly elected congressmen from entering their offices; vetoing a budget necessary to support the operation of agencies established by law; wholesale firing lawyers and creating a Youth Work Corps run by the military are all pure arbitrary acts of dictatorial madness."

The Speaker opened his attaché case and placed his notes on the podium.

"Abraham Lincoln, in the midst of a great Civil War, never treated congress with the insolence of our young President. Regretfully, with a majority of your elected representatives, I have been forced to draw up Articles of Impeachment. I will now read them to you.

- ***"l. Obstruction of justice by impounding money and unlawful firing of persons in government agencies.***
- ***"2. Abuse of power by unlawful declaration of National Emergency.***
- ***"3. Contempt of Congress, a body created by the Constitution."***

The Speaker allowed a moment of silence for his audience to digest the pronouncement.

Metzer then meticulously ticked off on his fingers the laws President Ericson had broken and how her actions took away the rights of the people to receive welfare, child care, food stamps, better education and universal medical care. He landed heavily on the emotional side of the issues fed him by the former Rhodes Scholar Harvey Hyman.

As Jodi attentively studied the Speaker's florid features, his shock of unruly white hair, the sweat dripping from his chin, she experienced a twinge of sympathy. 'Fat Albert' Metzer, like a dying breed, was savagely fighting his last stand. He desperately defended 30 years of a way of life in the U.S. Capitol. Like an old hard-line communist, he couldn't understand his disintegrating stewardship. He faulted a single evil usurper, not the generations of heavy handed-bureaucracy.

"One more thing," said the Speaker, "Presidents are supposed to enunciate solid foreign policies so that our people and the world will clearly understand our position."

Jodi attempted to interrupt but the Speaker's voice overpowered her.

"This President has no foreign policy," he hollered. "The nation is adrift."

As he gulped from a glass of water, Jodi tried again.

"My foreign policy is simple," she said, raising her voice to establish her position.

"We do not intend to invade or occupy any foreign country unless it threatens the United States. But that does not exclude slapping down or punishing with our Air Force and Navy any regime we determine threatens our trade or peace of the world."

She held up her hand to continue.

"We stand steady with a mighty big stick against 'crimes against humanity,' invasion of another sovereign country or acts threatening the security of the United States.

"While we may not choose to occupy, do not mistake that for weakness. Our punishment will fit the crime. Our retaliation could be specific or it could devastate and reduce an area to rubble and make governing impossible for dictators. People can no longer tolerate leadership which brings such reprisals."

"What about Haiti? Wasn't that a breach?"

"In our hemisphere, we reaffirm the Monroe Doctrine--so long neglected. Our back yard is to be regarded as an extension of our interests. Haiti was fast becoming a 'crimes against humanity' condition we addressed after Hitler. Furthermore, Haiti threatened our immigration and border control. It was overloading our support system."

"Another overseas adventure will produce more body bags," cried the Speaker.

"My new policy prevents body bags," said Jodi. "Politicians always fight future wars with past wars. If it were up to them we'd still use the English longbow. You forget we never invaded Japan with one American soldier. And neither the atomic bomb nor invasion was necessary. Japan was trying to make peace overtures through Russia but Stalin stalled. He wanted Japan to suffer greater punishment. Our incendiary bombs were already causing havoc and devastation to cities. The nation was turning into a fire storm."

"Another impeachable offense," the Speaker thundered, his hands quivering with anger as he pointed at Jodi, "President Ericson's turning the country into a NAZI training camp with a Youth Work Corps and Universal Conscription run by the military. Mussolini and Hitler started the same way."

Jodi patiently stood behind the podium rearranging her hair as she waited for the Speaker's applause to die down. When it became her turn she began in a quiet calm voice, hoping to lower the emotional intensity of the viewers. She was thankful for a large auditorium. Although a live audience had the potential for disaster, it gave her an instant barometric reading of the political weather..

"Pointing the finger of blame is how Hitler got started," said Jodi. "I'm sure the Speaker's Jewish friends will refresh Mr. Metzer's memory. So let's not worry so much about me, let's worry about the American people.

"I've been your executive barely one year. Speaker Metzer's been in congress 30 years. Three decades of taxes, burdensome laws, onerous regulations and hundreds of new agencies and incompetent programs. Congress and its leadership caused this mess we're in, not me.

"Crime, welfare, education, health care and their combined costs need solutions. Congress in the last twenty years has failed miserably in its responsibility to provide answers. It tried valiantly with the Republican landslide to change direction. But the effort fell far short of getting rid of the cancerous bureaucracy. Cancer isn't cured by excising warts.

"The Youth Work Corps under General Hardin is proving a howling success. Crime has dropped in half because kids are off the streets and into a disciplined environment. There's no need to spend $30 billion for prisons. Prison cells are for hardened criminals who commit felonies injuring people. And since idle youths were largely responsible for the dramatic jump in crime,Youth Work Camps have largely removed the basic cause.

"Teachers today teach. I'm told they're delighted with the leap in achieving educational goals. They no longer fight obstreperous classroom bullies and spend half their time maintaining order. They evict misfits and trouble makers. The Youth Corps picks them up, straightens them out and educates them in an environment which introduces pride and self-discipline."

Jodi continued. "Welfare rolls dropped. I've cut bloated staffs by two-thirds. Teenage pregnancies have fallen eighty percent. Idle males are not around. Single mothers, a huge burden to welfare, are no longer entitled to benefits unless they move back with parents or relatives. Even then they're required to enroll in some type of work effort before they can collect their checks.

"Hospital costs have skidded to an unbelievable low, and they will go lower. Youths with guns are off the streets. Hospital emergency rooms are not jammed with bleeding and dying victims. Medical costs associated with crime have plummeted."

President Jodi Ericson took a deep breath and surveyed her audience. The defense of the Youth Corps had touched on its accomplishments, not its military connections. She knew the deep seated resentment of the military by elements of the elite and professional bureaucrats. Since establishment of the Youth Work Corps, thousands of social workers had been cut from federal and state payrolls.

Marvin Pringle turned his questioning to Congressman Jesse Rowan.

"Reverend Rowan, what are your thoughts?"

The Chairman of the House Judiciary snorted his impatience.

"Not once has the President addressed the issue. Typical of her innumerable evasions of the law. Why should she not be impeached? She'd better attend that question or she won't be President much longer."

Jodi interrupted, her eyes cold with contempt.

"Our tax-and-spend friend serenely rides his Titanic course as it rushes towards the iceberg, never once thinking his own legislative body set up the collision. Congress, not the executive, needs impeachment. Congress over the past two decades has established a tax system which would be the envy of King Louis XIV. And it's headed for equal disaster."

"Trash," said a tight-lipped Rowan. "Pure hogwash. Your vaunted Work Camps snatched a disproportionate number of black students from their neighborhood roots. They belong home."

Jodi smiled wanly.

"First you wanted black youths bussed out of the neighborhoods. Then in a complete turn-around you wanted a return to their roots. Now you say black Americans can achieve full economic and political parity with white Americans and maintain a separate culture at the same time. It won't happen."

"Schools must accept black culture," insisted Rowan.

Jodi had a question.

"Culture which means a constant barrage of tormenting, high-fiving, loud-mouthing, profanity, loud music, disrespect and studied nonchalance?"

"Typical whitey bigotry," snarled Rowan, his face flushed with anger. "Schools got to accept ethnic differences."

"Instead they've accepted behavior which prohibits learning."

"Nonsense," said Reverend Rowan.

"Before Youth Work Camps, schools permitted conduct which simply stopped learning in its tracks. Misguided administrators and teachers allowed behavior to be dictated by adolescents."

"Segregated environments give black students better opportunities for leadership," said Reverend Rowan.

Jodi's tone was caustic.

"Most of our major cities are ruled by black mayors. Not one would have been elected without white support. Leadership on the basis of race is only one step above slavery. Only slaves need apply."

Reverend Jesse Rowan could barely keep the fury out of his voice.

"African-Americans have suffered 200 years of inferiority from whitey. Now you want to knock the props out from under their economic means of survival--welfare, food stamps and other social programs."

Said Jodi. "A true leader, who I'd like to see head a combined cabinet of Education, Labor, and Health and Human Resources, thought otherwise."

"Who?"

"Thomas Sowell. A man you should listen to. He questioned: is it 'acting white' to learn trigonometry? Even though trigonometry originated in Egypt, uses numbers from India, printed on paper originated in China? Were Europeans 'acting Indian' when they adopted Arabic numbers--learned from the Hindus--instead of cumbersome Roman numerals?"

"An Uncle Tom."

Jodi expressed surprise and pique. She responded contemptuously.

"You don't respect your own achievers? Sowell lays to rest arguments blacks haven't got what it takes when they apply themselves. General Powell and South Africa's Mandella prove such talk is nonsense. Like calling Jews incompetent, yet they produced an Einstein."

Reverend Jesse Rowan stammered in confusion.

"You're twisting words..."

"Not at all. Let me quote," Jodi said, tearing a piece of paper from her note-pad.

"Sowell said 'the whole history of the human race has been full of borrowings from one culture to another. Asians, Africans,

Europeans and others have all advanced this way...It would be hard to think of a more deadly formula for a whole generation to destroy itself than by glorifying ignorance and provincialism. The most fiendish racist could not come up with anything more devastating than such mass self-destruction....The talkers have banished the doers.'"

There was stunned silence. Marvin Pringle cleared his throat desperately trying to think of something to ease the tension.

Jodi Ericson stepped into the eye of the storm.

"You're fighting windmills, Reverend Rowan. Humans differ from chimpanzees in less than two percent of the three billion DNA bases in their genes. And the genetic differences within the human race are so minute as to be non-existent. Get with it. Better yet, get a work ethic and stop calling yourselves African-Americans. You're now Americans.

"There isn't a brain alive," said Jodi, "black, white, yellow or brown, that can't handle a thousand times more information than is poured into it. Geniuses throughout history have readily acclaimed that *perspiration not inspiration* was responsible for their success."

"You forget opportunity," insisted Rowan. "There's been damn little opportunity under slavery."

"Except you're no longer under slavery. You're leaning on a worn-out old crutch. Germans, Greeks, French and Italians were slaves in Roman galleys. The Irish suffered more deaths from starvation during the potato famine than blacks suffered during their entire history of slavery. Only a small percentage of plantation owners mistreated slaves. It wasn't in their best interests to injure their property. Slavery's an ugly, ungodly institution blatantly used by conquerors throughout history. Few peoples have escaped it. But now that's behind us. Today's a new world, Reverend Rowan. Time you entered it."

"Whites hate blacks."

"Not so. Most whites liked and loved blacks, many were nannies to their children. Agreed, the institution of slavery was an abomination which prevented whites from accepting blacks as equals, denied them their proper dignity or raised barriers to socializing. But whites never hated them. Blacks never endured the hatred of the whites in Boston against the Irish. Blacks were not denied food. Boston Irish were. Signs proliferated saying 'Irish need not apply.'"

"What's that to do with slavery?"

"Reminds us that many races suffered hardships. Quit whimpering about ancestor injustice. It was an unfortunate part of mankind's evolution. Let's get on with harmonious cooperation. I believe some of your strident leaders are no better than neo-Nazis in promoting ethnic diversity."

A furious Rowan took a step towards President Ericson.

"Calling me a Nazi?" Rowan raged.

Refusing to budge, Jodi spoke softly.

"Not at all. Only that sooner or later Americans will catch on the whole can differ from the sum of its parts. A fact known to mathematicians for a thousand years."

Marvin Pringle jumped between them, speaking hurriedly into Jodi's microphone.

"What do you mean, Madame President?"

"In Germany, under Hitler, no individual thought himself truly bad or evil. Initially, he merely sent his friend and neighbor to the Reichstag to represent his local interest. But once there, like congress, the neighbor insensibly became seduced into advocating great causes--great ideologies. These great causes spawned a Nazi bureaucracy of bigotry and oppression."

Seeking to lower rising temperatures, Monitor Pringle interrupted.

"We will now hear from Senator Mordecai Meyer," he said, signaling the Majority Leader to speak up.

Senator Meyer dropped his glasses to the end of his nose as he hastily rearranged his notes. Like a lost downhill racer, the debate had taken off in a wild unexpected direction. The victim had become the aggressor. Senator Meyer, an experienced campaigner, detected the shift in audience reaction since the Speaker's opening. Jodi had cleverly placed congress on trial instead of herself. Rowan's temper tantrum hadn't helped.

"Few in America," Senator Meyer said, "came to these shores to leave riches or kingdoms behind in the Old World. We were all poor, all rowing the same canoe--most likely stolen from the Indians. But no matter our past history the question before congress today which has to be answered is: 'should President Jodi Ericson be impeached?'

"To that question I regretfully answer yes. Why? Despite irrelevant arguments and considerable charm, she has broken the laws of our country. No citizen, President or otherwise, can do that with impunity."

"Laws passed by a decade of Imperial Congresses," injected Jodi.

Mordecai smiled. He was getting under the President's skin. Nothing could hurt her cause so much as an intemperate explosion.

"Laws passed by representatives elected by the people," he corrected.

Reverend Rowan, harboring resentment at the way he had been manhandled, hollered into the microphone.

"Our President is a bigot. Pure and simple."

Marvin Pringle signaled his engineers to deaden the mike. He feared Rowan might spew his rage onto the airways and give him trouble with the FCC.

Senator Meyer frowned. No sooner had he calmed the audience than the fiery Reverend raised its bile, producing a sympathetic reaction for Jodi.

"Impeachment was provided in our constitution as the only way of controlling a president from becoming a dictator," said the Majority Leader. "They were wise men. They'd seen enough of Russian Czars, European tyrants and George III."

Jodi determined to make one last pitch to the audience.

"The future lies in integration, not ethnic diversity," she said. "One planet. The view of astronauts does not display many diverse peoples. They see one planet. To integrate the planet we must begin like Christ. First bring together the disciples, then the nation--not by force but by cooperation, trade, language and the love of fellow man. Integrate the family, the community, the nation and our neighbors. From that base we can then hope to integrate the peoples of the planet."

Marvin Pringle glanced at the clock.

"Only minutes left, Madame President. Any quick additional remarks?"

"When George III put his heavy hand on our forefathers, they acted," Jodi said, stepping closer to the microphone. "Not by responding with more taxes to meet his demands--although that would have kept them out of trouble. No...They acted with a tough fundamental response of enormous courage. They started a revolution.

"Later, when our forefathers faced an incompetent Continental Congress which passed stupid laws, they again didn't try to patch up an inept body with shorter terms, bigger budgets or gerrymandered quotas. They got rid of the unworkable monstrosity and created a new constitution."

"One that's working fine," said the Speaker. "So why change something that ain't broke?"

Said Jodi. "Because we've cowardly backed into permanent gridlock. What's wrong with winning again? We quit before we finished the job in Iraq...we left Saddam Hussein in power to wreck havoc. We quit Somalia, turning power back to the drug lords. We fawn on an implacable foe in Cuba. Whatever happened to "unconditional surrender" terms which are necessary to forge a new beginning from the ground up?"

"Those were different times," snapped Metzer.

"Basic truths remain constant. After 200 years," Jodi said, raising her arms, "we need a new tea party."

Senator Meyer queried in nasty tones.

"Then what? A monarchy or dictator?"

"No. A different animal. One which stands tough as a sovereign nation. Yet with leaders who pick their way carefully through hazardous mountain passes."

"Like old goats," laughed Senator Meyer uproariously.

"Perhaps," smiled Jodi. "That's how they get old. But old goats can tame loud bleating sheep every time."

48. AFTERMATH

Jodi turned to Tony with a weak smile. She toyed with the zipper on her blue dressing gown. It was late evening following dinner after her TV debate.

"The bastards are determined to impeach me, come hell or high water," she said, picking at a loose thread over her bosom.

Tony watched silently. The Presidency turned normal people into tenacious animals--mothers in the wild savagely protecting their broods. Jodi now was one of the richest women in America yet she slaved day and night and worried over problems which didn't deserve the light of day. Whether congress impeached her or not wouldn't matter a tinker's damn to her economic welfare. Yet she was determined to push through her constitutional amendment even if it meant losing the Presidency. She worked longer hours and harder than a stevedore.

"There's nothing for me to do but keep going. Continue to take my case to the American people. Fight. Damn the torpedoes," Jodi said.

"One has to return to President Andy Johnson, Thaddeus Stevens and the civil war to find an opponent as malevolent as the

Speaker," said Tony, shaking his head. "I don't understand his obsession."

Jodi took a deep breath, exhaling the last dregs of air.

"I feel like a Viet Nam soldier torn from the grasp of a dead buddy and forced to continue on into a jungle of booby traps. I have no choice. I can't quit. I either make the grade or perish. Fate has taken over. I'm no longer in charge."

Jodi sat on the edge of the bed, her bosom heaving.

Tony watched the rise and fall of her breasts as she unconsciously loosened the zipper. Her breasts burst partially into view as she released the pressure of the material.

"Take it easy, Jodi. You're working yourself into a lather. General Hardin reports great success with all your initiatives."

"He's doing one hell of a job. But his feedback hardly slows the Speaker's vendetta."

"'Fat Albert' certainly didn't win today's TV war. He fumbled badly at times," said Tony.

"Wonder how the nation responded? My plea have any effect?"

"No way to know immediately," said Tony. "You asked viewers to jolt their congressmen, not the White House. We'll see the results by tomorrow."

Jodi knew Tony was trying desperately to raise her spirits. The debate had drained her dry. Though she hated comparisons she couldn't help reflecting on the different kinds of love offered by Guy and Tony. Guy's had been a consuming, demanding almost savage untamed animalistic love which drained and exhausted. Tony was gentle, understanding and considerate. Yet Tony had been the boxing savage in the ring and Guy the businessman. She wondered if she were a Dr. Jekyll and Mr. Hyde. Each fully commanded her passions.

Tony interrupted her thoughts.

"Jodi, you've got to select a Vice President. I realize this impeachment's distracted you. Yet as things now stand if you're convicted by the Senate the job goes to your hated enemy--the Speaker. I can't imagine anything worse."

Jodi got up and walked to the window, hands clasped nervously behind her.

"Nor I. That bastard as President would be worse than resurrecting Hitler for Germany. But you know I've tried."

"Yeah. I know. But the names you've floated have been anathema to your opponents. They're still bitter about losing their slate at the convention."

"I absolutely don't want anyone from congress. They're contaminated by years of sleazy self-indulgence." She hesitated, thinking of a few diehard friends. "With rare exceptions."

"Couldn't agree with you more."

Jodi swung around to face Tony with a half-smile.

"Tried my damnedest to get my old boss, Governor Aker, interested but he adamantly refused the job. Loves Nevada and wants to stay there. Plans to operate a large casino when his term is over."

"Yeah," grinned Tony. "He thinks Washington's got as many criminals in the capitol as in the prisons."

"Aker told me confirmation hearings were a Spanish Inquisition," said Jodi, "and he'd no intention of allowing partisan zealots to rip open his bowels to the public."

"What about your old friend from Nebraska? The guy who got you started?"

"Congressman Pace? I floated that idea but he quickly nixed it. He's from a state without much political clout plus his long-time friendship with me insures certain defeat."

"Too bad. My stint in Washington taught me Congressman Pace is a fine man."

"Politics, not character, is the name of the Washington game."

"One last suggestion," smiled Tony. "Senator Trudeau? She's not only above reproach but she introduced the two of us?"

"Got you there," smiled Jodi. "Thought of her first. At lunch one day I broached the idea. May laughed in my face."

"Why?"

"'Two females on the same ticket?' May nearly choked on her drink. 'A long-time staunch supporter of yours? Senate hearings would turn me into another Marie Antoinette. I've no wish to face that guillotine. Remember what they did to a highly qualified Bork? You've got to find your sacrificial lamb somewhere else.'"

"Fate of Vice Presidency hangs on more rational minds in the Senate," said Tony.

"Speaker calls me a tyrant. Am I?"

"Only in bed. And I'm not complaining."

Jodi smiled. "You originally thought I was an eggshell, easily broken."

"I've learned otherwise," Tony said, adding. "To my delight."

"Tyrant or not," Jodi's expression turned grim. "I don't remember legislative anarchy building any nation."

Tony studied the vibrant figure before him. A full matriarch in command of her femininity, Jodi bore only outward resemblance to the insecure, tentative, worried beauty from whom he had rented an office. She retained every bit of the fresh external youth, but one look deep into her eyes witnessed the change.

Tony swallowed hard as Jodi further loosened the zipper, completely exposing the deep cleavage between breasts. Her movement imprisoned his gaze and dried his lips until he was forced to wet them with his tongue. Jodi, consumed with releasing physical anger, noticed Tony's tense focus. She hesitated only a moment.

"Earth-shaking problems can wait? Right?"

"You are an earth-shaking problem," Tony said with a weak grin.

"Then we'd best take care of your libido or I'll never get your full attention." Jodi's lips moved with amusement, adding as she glanced at her watch, "I thought I took excellent care of that before breakfast. Do I have to service you three times a day?"

"A boxer learns it's better to stay in training at all times. Once you skip a few bouts you lose it."

Jodi zipped completely out of her gown in one swift movement and allowed it to drop to the floor.

She laughed at Tony's clumsy haste.

49. FBI & CIA COME CALLING

President Ericson's secretary called on the intercom.

"Mr. Keefe O'Hara and Mr. Fred Parkinson are here for their appointment. They have a Nikita Alexander and Leo Fedor with them."

"Please send them in," said Jodi, rising from her desk to greet the group as they were escorted into the Oval office. After arranging seating, the President selected a wing-backed chair near the windows. She greeted the two employees of Material Resources.

"Good to see you again. Last time was with my husband in London, right?"

"Yes Madam President," said Alexander. "I remember we discussed the monkey wrench Iran tried to throw into the dismantling process."

"Hasn't changed any since then, either," said Fedor. "Iran's worse than Saddam Hussein. Tests our tolerance to the outer limits."

The FBI Chief nodded agreement, his Irish face expressing his indignation.

"We're pretty damn sure Iran provided the assassins who killed your husband and President Roland," said O'Hara. "Even if they'd selected you as the target."

Jodi frowned. "You've told me about their mistaken target. So just what are you driving at, Mr. O'Hara?"

"We've uncovered heavy links to World Trade hoodlums and mullahs in New Jersey."

"Speaker Metzer's district?" Ali Rahman's evil threats fleetingly flashed across Jodi's mind. Was it possible?

"With the help of these gentlemen and Scotland Yard we're closing in on the details," said CIA Parkinson. "But it's been a struggle ferreting information across borders."

Said O'Hara. "The plot seems to have come together in Switzerland where the money was provided." .

"I understand the difficulty," said Jodi.

"The CIA's had employees working under the auspices of Mr. Alexander and Mr. Fedor," said Parkinson. "They've been extremely helpful."

"Your husband assigned us the task of giving complete cooperation to both the FBI and the CIA," said Alexander. "We continued that mandate after his assassination."

"I fervently hope you find the culprits. They destroyed the lives of two fine men." Jodi bit her lower lip white to keep the tears from falling.

"Scotland Yard thinks the idea originated outside Iran, Madame President," said Chief O'Hara. "Both Alexander and Fedor's operatives intercepted network messages flowing from Iran through Shiite Moslems in the Ukraine to an Ayatollah in New Jersey."

"The CIA obtained the information without their knowledge," added Parkinson. "They still use primitive communication devices."

"We know Iranian money was planted in Swiss banks to the credit of American Muslims," said O'Hara.

"Since they're numbered accounts we don't know who nor why," said Parkinson.

/

O'Hara rubbed the red stubble on his chin. In his haste to meet with President Ericson he'd neglected to shave.

"Scotland Yard, embarrassed and up to their ears in the investigation of the murder of a U.S. President on their soil, tells me the assassination definitely was a bungled job," said O'Hara.

"Why?" Jodi still had doubts.

"It brought increased vigilance into the flow of money and arms," said Parkinson. "Something Iran's Ayatollahs avoid like the plague."

"You're not making sense?"

"Hezbollah did the job. While Iran supports them with money and protection, the politicians often can't control the fanatics," said O'Hara.

Jodi agreed, heaving a sigh of frustration.

"Fanatics in any country operate in a world of their own."

"Treacherous animals. Often turn on their benefactors," declared O'Hara. "Like feeding alligators. Ignore their gluttony and they'll eat their handlers."

"Old-line communists are the ones selling arms under the table," said Alexander. "So why would Iran pay Americans?"

"Simple," said Jodi. "Iran wants to block the dismantling operation. Then kick the U.S. out of the Mediterranean area completely."

"That's our reading," spoke up Fedor. "Americans still don't understand Shiite Muslims. The threat facing us tomorrow is not a resurgent Russia. It's the millions of Muslim youths responding to the aggressive behavior of their mullahs."

"They're even a threat to Russia, as Yeltsin learned," said Alexander.

"I know," said Jodi. "Guy worried himself sick about Muslim ambition."

"Instead of feeding their bodies, the Ayatollahs feed idle minds with tripe," said Alexander. "Mullahs blame foreign devils for creating all hardships. They tell youths they're never responsible and that the Mullahs stand ready to solve their problems."

Fedor was unable to keep his eyes off Jodi. She definitely did not fit his idea of a head of state. Dealing with women in the Ukraine hadn't prepared him for a cover *girl who handled herself like an old-line cavalry officer.*

"Ayatollahs provide direction and assume responsibility for the young in Iran," Fedor said. "Shades of Khomeini or Hitler. In turn youths expect to be cared for. They follow the leader."

The FBI chief expressed his contempt. He had worked himself up from foot soldier to head of the agency.

"In the west youths today aren't much different. They demand goodies as inherited rights. You'd think some were children of royalty. They readily attach themselves to demi-gods who promise taxpayer handouts."

President Ericson smiled at O'Hara's passion. She understood his frustration at dealing with crime in hundreds of towns and cities across the country.

"We're reversing that with our Youth Work Corps," said Jodi. "And we're getting results."

O'Hara conceded the point with some reluctance.

"No question. Agents in the field are amazed. But their support is still tentative because they fear congress will cut off funds. Then they'll have a bunch of released kids to handle. Squawks from social bureaucrats are loud and insistent."

"Won't happen," said Jodi. "Not as long as I'm President."

A sudden silence followed her remark. The question hung on everyone's mind. How much longer could Jodi *Ericson remain President?*

Alexander brought matters into focus.

"Madame President, your latest edicts were right on target, but they sure raised hell with MRI's relations with the United Nations. Their snoops are giving us fits."

Fedor explained. "Your order blocking sales to Iran through any source caught UN cheaters with their pants down. Shipments intercepted at sea or in foreign ports waiting repackaging and trans-shipment caused chaos."

The CIA director rubbed the palms of his hands together nervously.

"Ambassadors at the United Nations somehow got the ear of Speaker Metzer. Several times he's met in Switzerland with Muslim leaders who demand greater leniency from your embargo. They want MRI and our forces in Ukraine placed under UN jurisdiction. Worries me."

"Under no circumstances will U.S. troops or commanders be subject to United Nations' command," said Jodi. "General 'Black Jack' Pershing established that principle in the first World War and I see no reason to change."

Said O'Hara with a grimace.

"President Clinton damn near gave our sovereignty over to the United Nations."

"We just wanted you to know the dismantling situation is volatile, Madame President," said Alexander. "Hate for the U.S. seems to be building, instead of diminishing. And the interference by some members of congress only roils the waters."

Jodi brushed the back of her hand across her forehead.

"There's only one thing a President can do to curb congressional interference," said Jodi with a resigned sigh. "Tell my treasury people to cut off their travel funds."

50. JODI IMPEACHED

Speaker Metzer, by the force of his position and arm-twisting patronage, finally achieved the impeachment of President Jodi Ericson. The battle for conviction was now moved to the Senate. Secret polls promised a squeaker. Sixty-seven votes were needed for conviction; thirty-four for acquittal.

Shades of the initial squeaker defeat under Senator Bob Dole of the constitutional amendment to balance the budget.

The Chairman of the House Judiciary Committee was ecstatic.

"I've chaired an historic gravestone for autocratic presidents." Reverend Rowan slapped his knee in high glee. "The House voted impeachment by a huge majority. The Senate's got to get the message."

"The House voted to save their ass...their jobs," cautioned Senator Meyer. "The Senate's got a few obstinate rich bastards who could care less. Some are about to retire, anyhow."

Howard Bloomberg, fired EPA administrator who owed his original appointment to the Speaker, spoke with a worried frown.

"Hope the process speeds up. Only reason I've not been removed from the job, the President's been too friggin' busy."

"President letting you hang in there until she appoints a replacement?" Senator Meyer was not fond of the Speaker's obsequious friend.

"Not exactly. She's not aware I've been dragging my feet in clearing out. Meanwhile she's planted an Inspector General in the agency and I can't shake the son-of-a-bitch."

The Speaker interrupted the rising animosity.

"I'm pushing with all possible speed. Let things linger and enemy forces always strengthen. Military axiom old as history."

The group was seated for lunch at a corner table in the PUB restaurant. They were celebrating the final vote for Impeachment of President Ericson taken that morning.

"You've got plenty chits in the senate," Mordecai reminded Metzer. "Money goodies to most states. You should have no trouble pulling a Lyndon Johnson and twisting arms."

Reverend Jesse Rowan was suddenly morose.

"My black caucus carries no weight with your uppity cohorts," he said, addressing Senator Meyer.

"So best keep a low profile," said Mordecai. "You've done a magnificent job to date. Don't spoil it by playing out of your ball park."

The Speaker nodded agreement.

"Right. The Senate's one black female sparks more animosity than a bull in a china shop. She's no Senator Brooks," he said, referring to a past beloved black Senator.

"Anything I can do?" Bloomberg wanted desperately to stay on the good side of the Speaker. He figured after President Ericson was kicked out of office the Speaker would restore him to his old job. And he understood the real reason for haste.

"We oust her quickly enough, that makes you President," Bloomberg said.

He displayed a satisfied smirk. Unfortunately he had mentioned the unmentionable. While the Speaker secretly kept that thought at the top of his agenda, he didn't like it cited in public.

"Keep that quiet," the Speaker growled. "We're impeaching a bad President, not promoting a substitute."

"Sure thing," said a chastened Bloomberg.

Senator Meyer barely suppressed a smile. He glanced at the red-faced environmentalist. No wonder the department was in trouble because of high-handed tactics. The Speaker took some crazies under his wing. His Administrative Assistant was another odd ball—a Shiite Muslim who consorted with Iranian crazies.

"Bloomberg's got a point," said Mordecai, wanting to gig the Speaker with a bit of needle. "Only thing standing between you and the Presidency is her selection of a Vice President."

"Never happen," said an annoyed Speaker. "She waited too long. Any appointment now would have to stand behind her Senate trial. Still I want it played low key."

"Makes sense," said Reverend Rowan. "In my hearings we learned President Andy Johnson's conviction failed partly because some senators focused on the next in line—Wade...a guy many couldn't tolerate."

"In thirty years a man makes enemies," agreed the Speaker. "Slights often are hidden under the skin. Let's stay away from me and hold the spotlight on Jodi Ericson. She's got plenty to account for."

"The Senate Judiciary Committee is nothing to fool around with," said Mordecai. "It's come a long way since the nauseating days of Biden, Metzenbaum, Simon and Kennedy. Today it occasionally demands proof."

"That we have in abundance," chortled Reverend Rowan. "President Ericson's broken so many laws my staff had trouble tracking them on a computer. Her trial's dead cinch for conviction."

51. TV BRAWL IN DALLAS

President Jodi Ericson slammed down *The Washington Post* so hard coffee spilled across the white Irish linen tablecloth. Tony Pellini said nothing as he sopped up the spill with a paper napkin and refilled her cup.

"They harp away at laws I've broken," exploded Jodi. "Abe Lincoln broke two to my one. The question is not who broke some technical law, but who broke faith with the American people."

"Hold steadfast to what you believe's right for America," said Tony, placing a hand gently on her shoulder. "Washington had his Valley Forge. Lincoln his copperhead press."

"I plan to stick to my guns," said Jodi. "No one's got a monopoly on brains."

"Particularly congress. Your call's good as any."

"Ninety percent of leaders who got into trouble did so because they backed away from their own instincts...got influenced by others," said Jodi.

"Truman refused to budge."

"But not Nixon. He always heard unfriendly footsteps, which he tried to hide instead of confronting."

"Reagan said his biggest mistake in eight years was believing congress spoke the truth," said Tony. "They promised $3 in cuts for every $1 in taxes."

"Congress hasn't kept a promise in 200 years, why would they start under Reagan." Jodi's lips curled in disgust. "They got away with another budget-buster."

"President Bush backed down when he knew better. His 'read my lips, no new taxes' will go down in history as the supreme cave-in of all time."

"Another wimp reaction which got him defeated," said Jodi. "I don't intend falling into that trap. People demand strength and the security it implies...A leader can be wrong and forgiven, but never weak."

"A truism forgotten by President Bush's 'read my lips' and unknown to President Clinton."

Jodi shook her head in amazement, her hair flowing back and forth about her neck.

"Probably no President in history so dramatically demonstrated that truism. Bush entered the war against Iraq with both Congress and the people skeptical. But he showed strength against indecision. His ratings zoomed to eighty percent. Then he turned wimp to an aggressive congress by abandoning his 'read-my-lips, no new taxes' pledge. Immediately, his high ratings plummeted. He got defeated by less than a 50 percent vote. Americans don't like wimps. Any future president should paste that reminder on his bathroom mirror."

Jodi lifted her face to the serene countenance of Tony. Through the recent weeks of grief he had been her unfailing champion.

"In your fights, how many times you get knocked down?"

"Four or five."

"Knocked out?"

"Once. A sidewinder blow. I wasn't paying attention."

"What happened on the knockdowns?"

"I got up to win the fights." Tony grinned. "I know what you're getting at."

What?"

"That I was in a confined weight class and you're in a no-holds brawl with monsters?"

Jodi stood up and kissed him on the cheek before striding to the windows overlooking the courtyard.

"You're getting so you read my mind. That could be dangerous."

"Or fantastic."

She gave him a sly look.

"Not the time for foolishness. Next you'll want to make love in the Oval office."

Tony hesitated, then smiled.

"Wonder if it's been done? We could be the first."

Jodi flashed him a look of disgust, then let loose a peal of laughter.

"I prefer bed. Besides, I doubt we'd be first. Midnight oil's no stranger to the Oval office."

Tony started to answer but Jodi turned her back and stared out the window. Tony could tell her banter was to cover her misery.

"Jesus was crucified for breaking narrow Hebrew laws but he established a religious philosophy which guided our entire western civilization for two thousand years," said Tony.

"Breaking bad laws is part of our history."

"Even the courts flip-flop every new generation."

"Our laws are supposed to express the will of the people. In the last three decades congress has passed laws thwarting that will and substituting their own Imperial grand designs."

"All requiring more taxes than the Czars of Russia."

The telephone rang.

It was General Hardin.

"They want you to speak in Dallas?"

Jodi nodded to Tony and punched the intercom button so he could hear.

She reminded General Hardin.

"Dallas hasn't been particularly hospitable to Presidents."

"That's changed," said the General. "You saved military bases all across Texas and the Youth Work Camps are new centers of attention for the local media."

Jodi laughed.

"Small wonder. Media kids expelled from schools landed in the camps."

"Nothing like involving your own kid to focus the parental mind," said General Hardin.

"Tony's standing here with me," Jodi said, not wanting General Hardin to think his conversation private. "I'm up to my ears in this impeachment mess."

"The city fathers and TV stations promise you a rousing reception," said Hardin. "Texas has a glorious military history. To them your impeachment is just another Alamo."

Tony watched Jodi as she stood by the telephone, looking tired, drawn and uncertain.

"You're right, General," Tony said. "Could be just the tonic the President needs. Away from Washington and talking to the American people on solutions instead of fighting ugly accusations."

Jodi bit off a hangnail on her little finger. She had neglected to have a manicure in the past two weeks. No doubt she needed a change of scenery.

"When? I'd need time for preparation."

"On their national TV hookup at noon one week from this Sunday," said General Hardin. "No additional preparation is required. I've got recent surveys from base commanders. Just keep hitting home runs on the positions you've taken."

Jodi smiled wryly.

"Sounds like you've got everything arranged?"

General Hardin's voice cracked over the speaker.

"Roger. Not a stone unturned."

Jodi wondered if Senator Trudeau had put the General up to taking such liberties with her schedule.

"Wouldn't it have been nice to consult me first?"

The General's abrupt answer brought her up short.

"No."

"Oh?"

"Senator Trudeau and I thought you'd nix the idea if it involved extensive preparations on your part."

President Jodi Ericson laughed.

"So May's fine hand engineered the scheme?"

"She got the Governor and Mayor interested."

"Okay, General. Make arrangements with my staff. I'll let them know how you and May browbeat me."

"Excellent, Madame President."

General Hardin hung up. Jodi turned to Tony.

"Seems I'm losing control of my own itinerary."

Tony shook his head.

"Hardly. Like all good military, General Hardin's deployed reconnaissance cavalry. He knows battles aren't won by a George McClellan sitting around the capitol."

* * *

President Ericson and Tony Pellini arrived in Dallas late Saturday evening. In the hotel they had dinner sent up to the suite, wanting to concentrate on Jodi's TV appearance and not face crowds.

Two waiters brought shocking bad news.

"Quite an event having two famous persons in town at the same time," said one.

"How's that?" questioned Tony idly, placing chairs at the small table.

"The President and Speaker together," the waiter said, lifting the cover over two Texas-sized steaks on a silver platter.

Tony stopped in his tracks, dropping one chair on his foot.

"You mean Speaker Metzer's in town?"

"On the third floor in this very hotel," said the waiter proudly.

President Ericson heard the news as she stood in the doorway drying her hands on a towel. She stared at Tony in dismay, biting her lower lip until the waiters departed.

"That bastard's got only one purpose being here. To sabotage my performance."

"He's got more tentacles than an octopus," said Tony. "Uses fellow members as spies. Plenty in Texas."

"Wonder how he expects to vandalize my talk?"

"Can't very well stand up and heckle from the audience," said Tony. "Monitor Jim Morrison runs a tight ship. Only his panel is allowed to participate."

"Maybe Fat Albert's big nose got the best of him? Perhaps he came to check my reception. If I flop he can charge full speed ahead on impeachment."

Tony shook his head dubiously.

"The Speaker's never been one to watch from the sidelines. He's got an active fiendish scheme of some sort."

* * *

Not until entering the studio the next day and meeting Senator May Trudeau and General Hardin did Jodi learn the distressful news.

"I did my damnedest to get Morrison to boot Metzer off the program," said a distraught Senator Trudeau, "but the Speaker wields too much power."

Jodi felt weak. Things could hardly get worse. She grabbed Tony by the arm.

"You mean they've put the Speaker on the same program?"

General Hardin watched Jodi turn pale and struggle to catch her breath.

"I'll be there with you, Madame President," he said, trying to bolster her confidence. "Fat Albert's said nothing new in the last 20 years. Our program's a winner."

Jodi gave him a look of gratitude, a bit ashamed at her momentary display. But the pressures of impeachment and the slings and arrows of the media over the past few weeks had cut deep. Constant bombardment weakens any fort.

"Thanks, Pete," was all she said, but her eyes said more. True friends show up under fire, not at the banquet table.

"Damn sneaky thing for Morrison to do," said Tony, "slip the Speaker in on the same program without telling the President."

Senator Trudeau slipped a supporting arm around Jodi's waist. Jodi had learned to love and treat May like a sister so the gesture was natural. Tony and May were now the only one's in her inner circle who had been with her in the beginning.

"Can't blame Morrison," said Senator Trudeau. "He fought Metzer's goons like hell. Head of studio overruled him only an hour ago. Speaker wields enormous power over TV stations and there was no light at the end of the tunnel for Jim."

"I understand," conceded Jodi. "I've mounted loco nags before. Some threw me, some didn't. We'll have to wait and see how I handle this one."

* * *

Guests were seated around an open semi-circle with TV cameras on both sides. Jim Morrison opened ceremonies by introducing participants and presenting background.

Speaker Metzer sat opposite President Ericson with the smile of a hyena lifting his face from a fresh kill. General Pete Hardin sat on Jodi's right with two local commentators completing the arrangement.

George Jamison, local TV commentator, spoke first.

"Today's national problems are crime, health care, welfare and education. How do you expect Youth Work Camps, Universal Conscription, and your proposed constitutional amendment to solve these? Seems they really have little to do with each other and will cost a bundle?"

Jodi looked beyond the stage into the audience. In the past she had wisely gained experience by insisting on appearing before live audiences. These encounters had taught her each interrogator carried his own brand of hidden bigotry into the questions. The truth was to be found in the reactions of the people.

"Doctors don't set a broken arm with a bandaid," she said. "Same with crime, welfare and education. They've got to be repaired at the break. Past methods of fixing everything with superficial bandaid social programs without treating the root won't work."

"Congress is now pouring money deep into root causes," said Jamison. "More prisons, more policemen, retraining workers..."

President Ericson interrupted impatiently.

"Congressional action to lock 'em up and throw away the key' always brings thunderous applause," sniffed Jodi. "It does less to reduce crime than increase the size of government."

Speaker Metzer leaned back in his chair with a tolerant smile, waiting for Jodi to alienate even more congressmen.

"Youth Work Camps are another escape route for juvenile criminals."

Said Jodi. "Most inmates clogging prisons aren't dangerous. Check forgers, small-time drug offenders, petty thieves and macho showoffs with nothing to do all day long end up in a court system with no other place to send them."

"Tell that to the small merchant operating out of a barred cage," Metzer said with a contemptuous dismissal wave of his hand.

"Stopping crime requires us to plant character and responsibility back into young people...to build conscience and backbone."

"Boot camps failed to reform 50 percent of the kids. Once out, they returned to their old tricks," said Metzer.

"Great," said Jodi. "You've just admitted 50 percent were reformed and sent back into society. And at incredible odds."

"Odds?" questioned Jamison.

"Boot camps were 90 to 120 days. Then they threw the kids back into society with no jobs, no money, no increased skills. One boy survived in a garbage dump for months before resorting again to dope and robbery."

"How will Work Camps do any better?"

Jodi felt like an ad man making a pitch to a disgruntled board of directors.

"Camps will keep the kids until they complete a high school education. They will attend classes in the morning, drill several hours each afternoon, perform productive labor until evening meals."

"And turned loose at night?" Jamison insisted on details.

Jodi smiled, speaking softly into the microphone and addressing the studio audience. She liked the mike attached to the lapel of her suit coat. It enabled her to modulate her tone to a conversational level without worrying about its position.

"Plus another hour of study before taps."

"No recreation time?"

"After an early five o'clock dinner."

"They'll still have a job problem," said the Speaker.

"General Hardin and I have a program in the works to solve that," said Jodi.

"Let's hear it." Conrad Whitmer questioned bluntly.

"We intend to place graduates in jobs."

"I suppose you think you can do better than the Department of Labor, Education, Health and Human Services and numerous other agencies we're pouring money into?" The Speaker's tone was derisive.

"Companies will get incentives." Jodi refused to stray from her purpose. She sensed the Speaker wanted to launch into the subject of her impeachment.

She explained. "Thirty percent wages will be paid the first year to companies hiring camp graduates, ten percent the second year. After that youths should be on their own and not forced to live in garbage dumps."

"How on earth can you force companies to hire...?" began Jamison.

"There's more," said Jodi.

"Oh?"

"Companies hiring these kids will have priority on government contracts. Five to fifteen percent on a percentage basis on the number employed will be credited to bids. Thus size will not provide any advantage." Jodi smoothed her skirt as the audience applauded and Morrison sought order.

"You'll still get many who can't find work," protested Metzer.

"Those who unsuccessfully try to find work will not be thrown out on the street like those from 90-day boot camps," said Jodi. "They may remain in the service for an additional year of study while concentrating with counselors to find work. An unemployed youth wandering the streets, getting into mischief and impregnating teenagers is far more expensive than additional training and education."

Conrad Whitmer had his own special agenda.

"The Department of Labor is spending billions on retraining and finding jobs. Isn't your program an overlap?"

"No. But theirs is. Retraining to do what? That Department wouldn't know a hammer from a nail. They shuffle papers and debate. Rhodes Scholars Clinton and his Secretary of Labor proposed cuts in corporate welfare. Reich said companies had too many welfare loop holes."

"He had a point," said Jamison.

"No. Reich was the one on welfare. Every dime of his salary was provided by taxpayers. But American companies earn every cent by production within the capitalistic system. His ideas came direct from Lenin—something he undoubtedly learned as a Rhodes Scholar."

Countered Jamison. "Yet even today the Labor Department's got its own training budget."

"I've already taken steps to eliminate the Department's entire retraining budget and move the money into Work Camps."

"Thousands of bureaucrats will holler bloody murder."

"They'll holler even louder when I eliminate thousands of needless jobs—theirs," said Jodi.

"How are Youth Work Camps going to help welfare or health care?" Jamison asked.

President Ericson turned to General Hardin. He had been intently following the verbal scuffle.

"General Hardin's had four months hands-on-experience. Let him tell you."

General Pete Hardin had just flown in from an extended survey of the camps.

"Results are hard to believe since we're still ironing out the bugs," he said.

"We're listening." Jamison was not interested in excuses.

The General stared hard at Jamison for a moment, then turned his back on him to face the audience.

"Available figures are fantastic. I've told my survey people to recheck them. Welfare costs are edging down while health care costs plummet like downhill skiers."

"Hard to believe."

Jamison was skeptical of government statistics for good reason. He'd nearly lost his job by swallowing false figures leaked to him when a cub reporter.

General Hardin explained.

"Youths aren't roaming the streets anymore. The nation's inner cities are clean. Major hospital costs came from emergency services—stabbings, gunshot wounds, gang wars, auto thefts, drugs, robberies and girl-friend brawls. They're now mostly gone."

"But welfare?"

"The President's new executive order requires teenage mothers to live at home. No separate status for teenagers. New pregnancies dropped 70 percent in four months."

"Not possible," declared the Speaker.

"True," said General Hardin, his jaw hardening at the denial. "Loitering young males aren't readily available."

Jodi sensed the nation carried a deeper problem than crime, welfare, education or health care. They were only surface symptoms. Democracy was in danger of not making it many more years. She determined there was no better place to air the danger than on national TV.

She said. "We can survive today's Four Horsemen of the Apocalypse—crime, welfare, education and health care—but our '*one nation indivisible*' can't endure if we don't get back to becoming one people, as our forefathers envisaged."

The Speaker's impatience finally erupted.

"They envisaged a government of laws, not an executive who ruled by whim."

Jodi ignored his ill-tempered remark.

"They sought freedom in a new nation. They fled ethnic hates and squabbles. They developed peaceful co-existence with harmonious neighbors whose ancestors had fought each other to the death. Our schools taught one language to bring them together. Our nation adopted one currency. Our forefathers sought an integrated country."

Jodi finished with a sigh. A deep hush fell on the stage and across the assembly. Were they listening or merely being polite?

"Like the cowardly sheriff who let the mob into his jail to lynch an innocent man, congress allows loud-mouthed ethnic and racial groups to promote their divisive agendas. Now this once-great democracy is headed back to tribalism. We're rapidly descending into what happened in Bosnia."

"Utter nonsense," cried the Speaker. "A democracy needs to listen to its tribal roots."

Jodi's lips curled in disdain.

"As Germany listened to its ethnic tribal roots?"

"As congress listens to its constituents," retorted Metzer.

"Districts gerrymandered to encourage ethnic diversity," said Jodi. "We'd be far better off creating brotherhood instead of friction."

"Wait and see, Madame President. Blacks, which your program hits hardest, will not sit still for your military Youth Camps. They will explode."

"Mr. Speaker," said Jodi. "Your remark categorizes blacks as being different humans than whites. The only difference is in the color of their skin. Youths of all colors, once involved, love the military."

"An insane remark."

"Blacks love the military. Today they comprise 20 percent of our volunteer force, nearly double their percentage in the overall population."

General Hardin leaned forward to remark.

"U.S. armed services proved the most effective instrument for affirmative action yet invented. Don't forget Arthur Ashe was a staunch supporter. He never abated his zealous pursuit of equality for his people."

"You bet," prompted Jodi. "Ashe believed in the military as part salvation for his people. He was a wise and educated man."

General Hardin added. "And General Powell rose to become head of the Joint Chiefs."

"In these camps General Hardin is providing the training and discipline needed to compete in the civilian market," Jodi said. "The 90 percent of our youth who don't belong in walled prisons still need to learn their actions have consequences. The ten percent true criminals who commit 80 percent of our violent crimes belong permanently behind walls. Prisons should be reserved for them."

"Camp costs will be enormous," said Metzer stubbornly.

General Hardin opened a notebook.

"Lawbreakers cost Americans an incredible $700 billion last year. And $122 billion of that was in added health care costs alone," he said. "Youth camps won't cost a fraction of that. Simple uniforms, plain food and good books are nowhere near as expensive as drugs and guns."

"The money would be better spent on education," said Metzer, his jowls shaking in anger at this young female fighting his every word. No previous president had so risked his wrath.

"Taxpayers have contributed to the highest per pupil costs in the world for some of the crummiest education. Thomas Sowell said it's as if 'incompetent carpenters decided to become brain surgeons.'"

"He disdains his own roots," said Metzer, his face turning beet red.

"No. He believed in America. Our early immigrants didn't rejoice in making the country multicultural. They dedicated themselves to the vision of a new nation—one language, one currency and goals possible in '*one nation, indivisible*.'"

It seemed that before she could catch her breath, the program was over.

To Jodi's surprise most questions from the panel had involved Youth Camps and Universal Military Conscription. Several times the Speaker, who the other panelists had relegated to a back seat, tried to insert comments about impeachment. Such intrusions were ignored. In the end Speaker Metzer furiously stomped out without a word, brushing past the extended hand of Monitor Jim Morrison.

Jodi worried as the unresponsive quiet and somber audience filed out of the auditorium. Were they for or against her? Uncertain of their true reaction, she hurriedly escaped by the rear exit.

52. SENATE VOTE TO CONVICT

Action on any vote involving a major national issue first produces arm-twisting backroom scenes before the vote reaches the floor. Frantic pressures, depending upon the severity of the issue, often escalate to the point of war. Friendships can be destroyed and alignments changed forever.

The Senate descended into a barroom brawl over the impeachment of President Jodi Ericson. Not since the civil war and the balanced budget squeaker had the Senate been so divided. President Nixon's case never reached the Senate, so final bloody lines were never drawn.

Past promises were dredged from forgotten wells. New vows pledged. Unused chits for critical aid were pulled out of deep pockets. And none were deeper than those of Speaker Metzer, who in thirty years service had pockets bulging with patronage chits.

Although not a member of the Senate which would vote on President Ericson's impeachment, Metzer patrolled Senate offices like a tiger cornering the kill. At some point in the past each Senator's perks and projects requiring money had originated in the

House. And he had been most generous. Now the Speaker was determined to call in each and every past favor—a vote for conviction.

The U.S. Senate never moves in haste. Each member must have his full say in debate. The backroom combat is as deadly as any on the floor. Spring came before that body was finally ready to vote. The decision was reached on a Thursday that the vote would take place at 11 A.M. the following Tuesday.

A final internal head count revealed conviction by five votes.

Senators queasy about throwing a U.S. President out of office insisted on one last caveat. President Jodi Ericson, now realizing conviction was certain, should be granted the opportunity to resign with dignity before the Tuesday vote. The Senate decided it should at least give her that much.

Majority Leader Mordecai Meyer was handed the chore of conveying the 'sense of the Senate' to the President.

In the Oval office Senator Meyer accepted President Ericson's offer of a seat. He wanted to make the message as painless as possible. The presence of General Pete Hardin and the ubiquitous Tony Pellini didn't make the matter easier. He had wanted to confront her alone. Jodi, however, knew the purpose of his visit.

The Majority Leader wasted no time on normal courtesies.

"Madame President, I have been dispatched by the Senate to inform you Tuesday at 11 o'clock that body will formally convict you on House charges of impeachment. A poll of members indicates your conviction will carry by five votes."

A drop of sweat fell from Mordecai's chin as he surveyed the silent trio surrounding him. Damn. And to think he'd had to fight tooth and nail for the Majority Leader's job. He blinked his eyes and plunged ahead when there was no response.

"The Senate will not bring this unpleasant matter to a vote if you will resign," he said. "We want to spare you and the country the pain of convicting a sitting President of the United States of 'high crimes and misdemeanors.'"

Tugging at his shirt collar with his forefinger he cleared his throat. He still faced deadly silence.

"Not only will it create an unparalleled bad precedent in this country but the effect upon foreign leaders is likely to be traumatic, perhaps injuring long-standing alliances."

The Majority Leader displayed one cheerful note of optimism. To President Ericson, who had been listening quietly without expression, it was the wrong note.

"Speaker Metzer stands ready and willing to step into the Presidency and smoothly carry on," Mordecai said. "We can arrange that at your resignation ceremony."

For a full minute the room was silent. Both Tony and the General waited for President Ericson to make the first move. Tension in the Oval office became almost unbearable.

Senator Meyer stood up to leave.

"I realize you're in a state of shock," he said. "You may phone me at my office when you decide on the details."

"No," Jodi said quietly, her soft voice under control. "Please sit down. I'm nowhere near shock. I've only been listening. Now it's your turn. There are some things I'd like you to report back to the Senate."

Senator Meyer sat down apprehensively. He'd delivered his message and wanted out. The faces of Tony and General Hardin were grim and foreboding.

"Senator, before I give my answer to your stark ultimatum I want you to hear out General Hardin."

She turned to the General.

"Pete, please give the Majority Leader the same report you just gave me."

"Certainly, Madame President," General Hardin said, moving his chair so that he looked directly into the face of Senator Meyer.

"In the nearly nine months since last September when the President initiated Youth Work Camps, a miracle has been happening in the schools and streets of America.

"The President instructed me to prepare old military installations as if we'd just suffered another Pearl Harbor. I did so, building a tight ship which ruffled many feathers. I drafted the best Army, Air Force, Navy and Marine Corps personnel."

"The military hasn't been that focused since Desert Storm," said Tony. "Bases hum with activity."

General Hardin continued. "State and local enterprises were given first access to the enormous purchasing power of bases within their vicinity. After a shake down of three months, citizens were invited and encouraged to visit camps. Parades regularly are held on local and civic events when the kids get to strut their stuff."

"Boot camps got failing grades all over the nation," grumbled Senator Meyer. "Drill instructors engaged sadistic practices. They dehumanized the young. Then when the kids got out, they reverted to their own innate inhumane behavior."

"You're quite right," said Jodi. "I abolished the 90 and 120-day boot camps. The Youth Work Corps sponsors education, pride and civil behavior."

General Hardin's answer was terse.

"There's no relationship to boot camps. Youth Camps are run by the active military. Abuse doesn't exist. Not under my watch. We're education and achievement oriented."

"You've obviously never visited a camp," said Tony. His irritation with the negative attitude of the Senator showing in his clipped speech.

"Well..."

"Have you?" pressed Tony.

"Frankly, I haven't had the time," said Mordecai. "Impeachment proceedings..."

Tony interrupted.

"Otherwise you'd know visitors are welcomed every Sunday. Families participate in using base facilities: swimming pools, picnic areas, golf courses, baseball and football diamonds, and basketball in the gymnasiums. Not an atmosphere for abuse."

"On Saturdays units compete fiercely to march with the flag at the head of the parade," said General Hardin.

President Ericson staunchly supported the job General Hardin was doing.

"If you'd check, Senator, you'd find crime in the inner cities is now confined to hardened adult criminals. Juvenile crime rates have plummeted 70 percent. Education in the schools again has become a joy to teachers. Disciplinary problems have been kicked out the door."

"Youths aren't stupid," said the General. "They reacted immediately once they learned expulsion into work camps awaited troublemakers."

Tony, who had been assigned the job of tracking costs, added his findings.

"Health costs, with emergency rooms reduced to automobile accidents and normal emergencies, have fallen over $1 billion."

"Better yet," said the General. "Drug tests are mandatory every 30 days. And we apply no penalties, so there's no stigma. Instead, we assign the victims to doctors to effect cures through controlled rehabilitation programs. Granted, the military has the distinct advantage of being able to control behavior."

After a brief silence, President Ericson rose from behind her desk.

"Senator Meyer, now you may go. I hope you will have the courtesy and good sense to acquaint your reluctant colleagues of the things which my National Emergency has accomplished."

"I'll take it under advisement," said an angry Senator Meyer. He'd come to deliver a message and got a lecture.

"But the Senate intends to vote Tuesday come hell or high water," he added. "They've debated this impeachment many long months and won't budge at this late date."

"TV, radio, and newspapers will get the message you've just received," said Jodi.

"You don't intend to resign?"

"Absolutely not.

"Conviction is certain."

"Not when the public hears the full story."

"The public doesn't vote. The Senate does. And the Senate has already made up its mind," said Senator Meyer, his hand on the door.

"Perhaps five voters will listen to their constituents," said an unmoved President Ericson.

"Be wise. Save yourself the humiliation," said Mordecai with a tight smile. "Tuesday you will be convicted and expelled from office."

"This is Thursday," said Jodi. "In four days Napoleon lost Waterloo, Spain the Armada, France control of the seas at the battle of Trafalgar, and Darius lost Persia to Alexander at Arbela. In four days the world turns four times."

Jodi expressed a bravado she didn't feel. But like any cornered animal, escape lay only in attack. She waited until Senator Meyer left the room, turning with a spirited front to Tony and General Hardin.

"Looks like we've got four days to overthrow a Capitol fortress," she said. "Will we end up another Wellington or Napoleon?"

53. AN ASSASSIN STRIKES

"Before I leave office this is one more thing I've accomplished," President Ericson said in signing a hard-fought bill at a news gathering in the White House.

Speaker Metzer stood to one side beaming hatred at what he was certain would be Jodi's final act as President. The ceremony followed by less than one hour Senator Mordecai Meyer's message on impeachment.

"This should insure that Iran gets not one dime of arms or weapons of any kind over which the U.S. has supervision," said Jodi, signing with a flourish.

"Let's hope it's only the first step," said Senator May Trudeau.

"Materials Resources International will see to it that none of the weapons of mass destruction in Russia, Ukraine, Belarus and Kazakhstan ever reaches Iran," said Jodi. "We've allocated over $1 billion for the destruction of such weapons and we don't want Iran sneaking any across the border."

"Well done, Madame President," said Senator Trudeau.

"Iran is the new 'evil empire' which threatens nations of good will," said Jodi, handing a pen to each of those standing nearby.

She continued speaking as she wrote, casting a smile in the direction of the seething Metzer.

"Iran's history of supporting terrorism, torture and murder under religious fanaticism sets it apart from a world increasingly demanding humanitarian principles of behavior. Iran's Ayatollahs and Mullahs are creatures from time past: bigots and purveyors of hate towards fellow men."

Only nervous light applause greeted Jodi's statement. Several members glanced at the Speaker, his face filled with rage. Metzer had fought a losing battle to lift all restraints against Iran. Mullahs were his strongest means of financial support. His New Jersey district had become the most populous site of Shiite Muslims in the United States. But the rest of the country did not think like his constituents.

As the gathering started to break up, President Ericson rose from her chair. Her jaw jutted belligerently.

"Now for my major announcement."

The group halted like a snared rabbit. Here comes the big news. Speaker Metzer leaned forward with a half-smile. Senator Mordecai Meyer's message finally had got through to President Ericson. She was about to announce her resignation. An erie quiet fell over the assemblage as they waited with baited breath for the historical proclamation.

Jodi smiled.

"Today I'm nominating General Pete Hardin to be my Vice President."

Bewilderment and consternation were the sole reactions.

"I'm sending the appointment to congress first thing in the morning," President Ericson said, before turning on her heels and exiting the stunned group.

* * *

"That ought to temper the Speaker's ardor," said Tony Pellini walking with the President back to the Oval Office.

"I think not," said Jodi.

"No? You just reduced him from a dreaded rat to a mouse. If Pete gets in before you leave, Metzer's lofty ambition to become President hits the garbage pail."

"Thomas Fuller once said, 'though an enemy seems a mouse, watch him like a lion.'"

"One question?"

"Shoot?"

"Why this last-minute ploy?"

"Several reasons," said Jodi. "I sensed it might come down to this. Speaker Metzer, despite his powers in the House, does not have too many friends in the Senate. He's bullied and browbeat members with numerous threats in order to convict me. This will throw a monkey wrench into his timetable."

"I don't see how. The vote is scheduled for Tuesday noon."

"I suspect a majority of the Senate will insist that a Vice President be in place before the vote."

"You know that for sure?"

"No. But Senator Trudeau does. She conducted a secret poll of her fellow Senators."

Tony wiped his forehead.

"Holy jumping Jesus. I thought the KGB was dead."

Jodi laughed, placing an affectionate hand on Tony's shoulder.

"May's prepared to introduce a motion to postpone the vote on impeachment the moment my formal announcement on General Hardin reaches the Hill."

"That why you kept stalling on appointing a Vice President?"

"Partly. You know I originally wanted May, but she'd have no part of the job."

"Two females running our government? A bit much."

"Watch it," grinned Jodi. "You're treading on quicksand."

"And the General?"

"Prior to now I wanted him away from politics. He needed to handle the operation of the Youth Camps under the military."

"You trusted no one else?"

"Correct. The bureaucracy would manhandle, delay and dominate anyone but a strong military figure."

Tony answered his own question.

"Reason you jumped him over the old guard to make him Chief?"

"That, and Senator Trudeau's support. She's known him since high school."

"Yeah. She's the only one who took your side when the Speaker ran you out of Washington."

"Not so," said Jodi, reaching over and taking Tony's hand. "You were there."

Tony, reddening to the base of his ears, protested.

"I meant in congress."

At the end of the hall, President Ericson turned tired eyes to her friend.

"Tony, I'm exhausted. I need a nap. Please tell the cook I'm skipping dinner. Perhaps when I'm refreshed the two of us can take a walk in the Rose Garden later and discuss plans for the next three days. Okay?"

"Of course. An hour's rest does wonders. Call me when you need me."

* * *

Dusk was settling over the rear lawn as President Ericson and Tony strolled around the oval track behind the White House. Jodi was her radiant, cheerful self again.

Tony, walking quietly longside, studied her incredible features as she concentrated on peeling an oversized orange with her fingernails. After each rebound from tragedy she appeared more beautiful. Leonardo Da Vinci said 'In life beauty perishes, but not in art.' A gifted artist was needed to capture this moment.

Jodi wore a white turtle-neck soft cotton sweater which rested snugly up against a chin tightly sculptured as if there were no marble for excess. Tanned skin stretched over a face without a wrinkle. Her mouth still was a mystery to Tony. Not wide, it was deep and sensuous with full fleshy lips which stirred the senses in the pit of his stomach. Blue-green eyes contrasted sharply with the copper skin and high cheek bones. A rodeo rider's broad shoulders supported the falling soft waves of coal black hair. Tony chuckled. Jodi was a genetic fusion as intricate as an atomic bomb.

"Something tickle your fancy?" said Jodi, stuffing an orange peeling into a side pocket to prevent littering.

"Just wondering what the good Lord had in mind when he put you together."

"You think he missed a few parts?"

"No. Just amazed that he found them all. They're not in the normal parts' department."

"No matter. They function with the same agony and joy as yours."

"I doubt it."

Jodi stopped her pacing.

"Meaning what? That I'm some Mongol freak?"

Tony reacted as if he'd been slapped. Instantly he realized the pressure she'd been suppressing was getting to Jodi.

"You know such thoughts never entered my mind. I've adored you through the years."

"Then what?"

"My dream would be to chuck this pain and run away to some deserted island. I'm selfish. In all these years I've never been able to get enough of you. Always I'm sharing ...."

Jodi reached over and patted Tony on the hand.

"Your time will come. Be patient..."

At that moment a rifle cracked and Jodi slumped into Tony's arms. Feverishly he lowered her into his lap and covered her with his body. Within seconds the Secret Service was by her side calling for the White House ambulance.

Jodi, her face ashen as the blood drained out from her body, retained consciousness. She looked up into Tony's distraught eyes.

"I love you, Tony," she managed. Then as the blood continued to flow, Jodi collapsed.

54. THE HUNT

The next 12 hours President Jodi Ericson was on life support in critical condition from loss of blood. No vital organs had been hit but a shoulder artery near the armpit had been sliced which paramedics had a difficult time stanching before reaching the hospital.

Tony Pellini never left her side. Senator May Trudeau was in and out a dozen times and Governor Aker arrived early Friday morning from Arizona. People across the nation were outraged at another assassination attempt on a U.S. President.

The surgeon who repaired the artery faced the media early Friday morning. He had spent most of the night in the operating room.

"The President's not out of the woods yet," he said. "But she's got a strong body and unbelievable stamina—also a generous measure of courage. She's fully conscious. At this point we can only hope her heart stands the strain."

Immediately following the hit on the President, Secret Service agents pin-pointed an upper floor of the Treasury Building as the site occupied by the assassin. From there the gunman had a clear view of the White House back lawn. The FBI swarmed the streets and closed off the building until all occupants were accounted for.

"Course a smart assassin would have cleared the building before we got here," grumbled FBI Chief Keefe O'Hara.

Guards at every entrance were grilled meticulously. The shooting took place at 6:05 p.m. while there was still light. The building was nearly empty by then. Chief O'Hara reasoned the hit-man would have entered when activity was high, before the evening exodus. That meant he must have entered before five o'clock. The guards changed at five-thirty.

Since the assassination attack occurred just after six p.m., all guards on duty at that time were rounded up in the Treasury hallway. Only four of the guards had been stationed at exits to the building. All other exits were closed at six.

FBI Chief O'Hara was particularly interested in the exit guards.

"I need to know the names or descriptions of everyone who left the building through your exits," he said. "Obviously the culprit entered before you came on duty, but he had to leave during your watch. Try to remember any briefcases, suitcases, and suspicious or nervous behavior."

The sergeant spoke up.

"Leaving with a weapon was impossible," he said. "We've metal detectors which all packages must go through."

"Unless it's a Senator or well known staffer," said O'Hara. "They're waved around the electronic surveillance? Right?"

The sergeant's face reddened.

"Yessir. When it's crowded that sometimes happens." He straightened his shoulders. "But at that hour it's never crowded."

Chief O'Hara noticed the sergeant's embarrassment and his suddenly defensive posture. The worst thing that can happen on an investigation is pushing someone into a protective shell. They clam up when they think they might be disciplined or fired for neglect or negligence.

"We've no wish to embarrass anyone," said O'Hara gently. "We all make mistakes. I assure you no one will be censured or reprimanded. Instead those who provide salient clues will be cited for their cooperation. Do you understand?"

"Yessir."

"Now. A criminal tried to assassinate our President. That's a felony of the worst order so think hard. Were there any unusual circumstances."

A youthful guard raised his hand.

"Speak out," said O'Hara.

"A big man with a musical instrument tried to leave about 6:10." The youth hesitated. "He tried to rush around the detector."

"Yes?" O'Hara was immediately alert.

"When I insisted he place his case on the inspection belt he flashed his pass and started on. I stopped him again.

"Don't you know me? He practically hollered in my face."

The youth looked at his sergeant questioningly.

"Go on," said O'Hara.

"No I don't, I told him. I've only been on this assignment for three weeks and I know few people. I thought for a moment he might hit me. He was huge."

"I'm with the Speaker, he said, flashing his badge and again starting forward. I stepped back and blocked his way. I've my orders, sir. No exceptions, I told him.

"With that he stormed back to the elevator. That was the last I saw of him. He must have exited through another door."

A second guard spoke up.

"A big man hurried though my station about that time but he had no case or package of any kind."

"Means he had to quickly stash the weapon and it's still somewhere in the building. My men will just have to find it."

Chief O'Hara thanked the guards for their cooperation and promised the younger guard would get a personal commendation for sticking to rules.

Agents were immediately dispatched to pick up the Speaker's Administrative Assistant. By this time several descriptions left no doubt who the man was.

A search of the building found the weapon an hour later in the trash deposit where the assassin obviously hoped it would be discarded along with the mountain of debris dumped each day.

The weapon was a .30-caliber rifle with the scope and barrel disassembled and all wrapped together.

“A sloppy job,” said Chief O’Hara, examining the rifle. “Obviously done in anger and in a highly agitated and emotional state.”

“Nothing professional we need to worry about, Chief, that’s for sure,” said his assistant, gingerly taking the weapon and handing it to his lab assistant.

“Fingerprints all over the rifle, the case, the scope. He certainly didn’t intend for anyone to find it,” said O’Hara.

“An all points bulletin’s been put out on his car, which our agents found missing. He didn’t manage to grab much from his apartment,” said the assistant.

“Panic blazes an elephant’s trail,” said O’Hara. “Probably heading home to New Jersey. Highway patrol will pick him up.”

FBI Chief O’Hara proved right on the money. The highway patrol picked up Ali ‘Knuckles’ Rahman just before the New Jersey state line.

* * *

Friday morning Ali Rahman stood before Chief O’Hara in the interrogation room of the FBI building. Ali had been returned to jail in Washington. Efforts to reach his boss, Speaker Metzer, had thus far been unsuccessful.

O’Hara studied the typed and signed confession before him. He glanced at his wrist watch. Eleven hundred. He was getting hungry. It had been a long and sleepless night.

“These are mighty serious charges to levy against a high public figure. Working for a plea bargain by dragging down a big shot?”

Ali’s mind didn’t fit his bulk.

“I was promised a plea bargain if I told the truth. That’s what I’m doing.”

“We’ll need corroborating proof.”

"I've got bank account numbers in Switzerland. He had me receive the money and make deposits for him many times," Ali said, perspiration dripping from his forehead.

"Against making the trips himself?"

"Too many jaunts could easily have caused suspicion," said Ali. "Anyhow, trace the gun. It belonged to him."

"You fired it."

"He gave it to me late yesterday. Right after the President signed the order banning any kind of sales to Iran."

Chief O'Hara stroked his chin, a slight smile appearing at the corners of his mouth.

"It's common knowledge you and the Speaker harbored no love for the President?"

"I hated her. She appears beautiful and innocent to the public but underneath she's a savage."

"So the idea of shooting the President wasn't abhorrent to you?"

"They shoot marauding tigers, don't they?"

"You attended the ceremony?"

"I was in the anteroom."

"Then what happened?"

"Speaker Metzer was furious. He stormed out, grabbing me by the arm and catching a cab to his apartment."

"Where he gave you the gun?"

"And a promise of $100,000 to do the job immediately. He said he was leaving and would meet me at my father's Mosque in New Jersey where we could get witnesses to swear we were there at the time of the shooting."

"When news you bungled the job reached him he must have changed plans?"

"Damn new kid at the door didn't let me pass," complained Ali, as if his troubles were all the fault of the guard.

"But you had no trouble entering the Treasury building earlier with a dismantled rifle under your arm?"

"Trombone case."

"Even so..."

"The day guard knew me."

"Anything on this written statement you wish to change?" Chief O'Hara knew charges of pressure could mess up a prosecution.

"No sir." He thought for a minute. "Except it was Metzer's idea and I'm not going to sit still as his fall guy."

Chief O'Hara nodded to the two armed agents standing beside Ali Rahman. They took him by the arms and led him from the room.

O'Hara promptly picked up the phone to reach Tony Pellini at the hospital.

"How's the President doing?" His concern was real. She had been a good friend to the FBI.

"Coming along fine," said Tony. "She's off the critical list. Now it's only a matter of getting her strength back."

"She gonna be okay?"

"Roger. She's had enough transfusions to stand an elephant on its feet."

"Ali Rahman spilled his guts last night and this morning."

"Any new developments?" Tony had kept in sporadic touch with the FBI.

"Ali would sell his mother to plea bargain. He insisted Speaker Metzer was behind the whole scheme."

"That fat bastard," Tony cried. "He'd kill an angel to satisfy his greed and ambition."

"We found out one thing more," said O'Hara. "Ali spilled his guts clean as a plucked chicken."

"Yeah?"

"Metzer was behind the assassination of President Roland and Guy Ericson."

"My god. Why?"

Said O'Hara, switching the phone to his other hand.

"A botched job. Metzer fingered Jodi but with their usual efficiency Hezbollah used amateurs."

"Cheaper and more expendable," snorted Tony. "But two men mistaken for a beautiful woman? Hardly possible?"

"Nervous amateur fanatics," said O'Hara. "Guy Ericson wore a purple bathrobe which threw them off base. The minute they saw purple they began firing."

Said Tony. "Dirty son-of-a-bitch. Metzer never let go his hate and humiliation. Where's he now?"

"He's somehow slipped through our net. We can't find him."

"Must be some net," growled Tony. "Let a whale like 'Fat Albert' slide away."

"Had an eight-hour head start," said O'Hara, understanding Tony's frustration. "But we'll snare him sooner or later. Tough to hide a blimp like Metzer."

* * *

By late afternoon Friday the media had the whole story. Ali Rahman's assassination attempt from the Treasury building across from the White House; President Jodi Ericson's battle against death; the Speaker's depravity and escape; and her challenge to the Senate by her refusal to resign. She had demanded that votes against her be posted and recorded for the nation to see.

Saturday and Sunday were field days for the media. The hunt for Speaker Metzer took on international proportions. A national wave of sympathy swept across the country for their President.

Those senators who had been vocal buddies of the Speaker supporting conviction suddenly disappeared into the woodwork.

The FBI learned the Speaker had used his ample start to full advantage. He'd caught the Concord from New York to Paris, where the French gendarmes arrived at the Orly airport twenty minutes too late.

"Speed was more important than camouflage at this point," said O'Hara.

Since the FBI Chief believed Metzer would attempt to pick up his money in Switzerland, he arranged for the Swiss gendarmes to cover the bank when it opened Monday morning. Ali Rahman readily had furnished the name of the bank.

The trap worked. Speaker Metzer was nabbed as he emerged from the bank with a suitcase full of money. The Swiss found a ticket to Tehran in his billfold.

Chief O'Hara got on the phone with his counterpart in Zurich.

"Will extradition be a problem?"

"Not in this case," said Inspector Giglund in perfect English. "We don't wish to become embroiled concerning a Speaker in the United States Congress."

"Good. I'll have two agents leave today." O'Hara was curious. "Any trouble picking him up?"

"None. I had men inside the bank." Inspector Giglund laughed. "His disguise was rather pathetic."

"How?"

"Wore Arab clothes, heavy black glasses and a mustache. But couldn't hide his bulk, which wouldn't have made any difference because we were watching for large withdrawals."

"Yeah," said Chief O'Hara. "Like putting a mustache on a walrus. Doesn't disguise the animal much."

"I've kept him separate from our other prisoners, so be sure your agents have the power to pay the bills."

"Oh?"

"My boss insists that's done before he signs extradition papers."

"What bills?"

"Hotel, for one. I've got him in a hotel room guarded by two men. Didn't want to process all the paper work normally required to send a man to our prison."

The FBI Chief smiled to himself. Once a Swiss, always a Swiss. When it came to money, they didn't miss a trick.

"No problem. I'll send an agent with the power to settle accounts," said O'Hara. "We appreciate your cooperation on this matter."

No sooner did Senator May Trudeau hear about the capture of Speaker Metzer than she rushed to stop the trial against President Ericson.

“Let those who would vouch for the integrity of Speaker Metzer now come forward and stand before this body and the American people,” she said from the floor of the Senate. She waited in silence for a full minute.

“I see there are no takers despite the fact a few days ago many of you clamored for that privilege—or was it the perks and goodies Speaker Metzer promised in appropriations?

“I move we vote at once to drop any further action against our President, who nearly suffered death because of her determination to improve the lot of our people. Do I hear a second?”

A chorus of seconds filled the Senate chamber. In short order the vote to drop charges carried 70 to 12 with 18 in absentia.

55. NEWS FROM SWITZERLAND

Senator May Trudeau sat by the bedside of the President attempting to cheer her up by telling her the good news of the dropped charges.

Still weak but spry and alert, Jodi reached for May's hand as Tony and General Hardin stood by.

"Perhaps after all these years those doubting Thomas' will believe Ali's and the Speaker's mistreatment of me years ago."

"May and I believed from day one," protested Tony.

"Like Damon trusted Pythias," smiled May.

Jodi looked into Tony's dark eyes and saw a frightening devotion. Had it been there all along and she too busy to recognize it? Had it taken a hospital stay for her to recognize that day-to-day intimacies were the most important things life had to offer?

"Dear Tony," she said. "When I thought I was dying my thoughts were of you. I feared you would be miserable without me. At that moment I knew I truly loved you. There were no others—ghosts or real."

Tony's eyes brimmed with tears.

"All I want is for you to get up off the canvas and start punching again," he said, wiping his eyes with the back of his hand.

Jodi abruptly changed the subject

"The doctor says I'm fine. No lifting anything heavier than a book and light foods until the artery is completely healed. Otherwise, proceed as usual."

"But with caution," said May.

The telephone at Jodi's bedside rang.

"Must be important. I told them to screen all calls."

Tony picked up the telephone.

"Tony Pellini here." After listening and nodding, he finally said. "Okay, Chief, I'll tell her."

He turned to Jodi.

"That was O'Hara. Your lifelong enemy just committed suicide. Speaker Metzer jumped out of an eight-floor window in Zurich."

The news momentarily stunned the group. Senator Trudeau reflected.

"Thirty years in congress? What a shame. A power base so painstakingly built creates panic when threatened. This President threatened his congressional power like no other," said May.

General Hardin commented.

"History is full of incidents where men refused to let go of power and privilege until they were physically dethroned."

"Hitler was one," said May. "He too committed suicide."

Jodi turned her gaze on General Pete Hardin who had been sitting quietly in the background. Rarely were his eyes off May Trudeau.

"General Hardin," she said. "I think it's time we got to work and pushed through your nomination for Vice President. If my amendment gains momentum in the states, I've no intention of campaigning for a second term. But I want to leave behind someone with the strength and character to give this country back to the people."

"You've picked the right man," beamed Senator Trudeau.

"When I think of how close Metzer came to being President, it frightens me," said Jodi.

"You were lucky he didn't hire a professional killer if he was that determined," said Tony.

"Emotions got the best of him," said General Hardin. "Same with Ali. Even an expert marksman can't hit the bullseye when he's seething with rage. Reason on the range we first teach firing by the numbers. Emphasizes control."

Jodi winced with pain as she moved. She made a feeble attempt at humor.

"Least a professional wouldn't have hit the shoulder," she said, turning to Senator Trudeau.

"May. It's up to you to push the nomination of Pete. At last they've got to realize how vital it is to have a sitting Vice President."

Senator Trudeau smiled. She refrained from reminding Jodi that she had been the one who stalled so long on the selection. Instead she agreed.

"Pete's nomination should clear quickly. I've already tested the waters. Your short-time inability to govern was frightening."

"About the camps? Are they ready to be released to a second in command?"

"I'll still oversee them as my number one priority," said General Hardin. "I've been brutally selective of camp commanders. Each one's been personally interviewed at length. They're 100 percent in accord with your objectives, Madame President."

"He's inspected every camp," volunteered May. "I accompanied him on several trips. Pete's penchant for detailed compliance is sometimes frightening."

"I consider orders are either obeyed in total or they're not obeyed at all," said Pete. "I know of no war won by an undisciplined force."

"The dramatic turn-around in crime, school discipline, welfare roles and health costs has penetrated a lot of diehard minds in the senate," said May.

"Helped along by constituent pressure," said General Hardin. "I've seen the change in those visiting camps—both parents and citizens. Nine months has made a difference. They've regained as much pride in their kids as the youths have gained in themselves."

"Discipline and achievement will do that," said Jodi quietly.

General Hardin added an observation gained from years in the military.

"Never underestimate the fertile soil in young minds."

56. BACK TO WORK

The following Friday President Ericson was back in the Oval Office. In the preceding week a host of things had happened, demonstrating how fast congress can move if motivated.

Tony sat in a chair near the windows of the Oval Office as Jodi tackled the mountain of paper work. He had postponed his return to Las Vegas by establishing a Materials Resource office in Washington from which he could direct affairs.

"General Hardin's confirmation breezed through like it had wings," said Tony. "In three days. Unbelievable."

"Our next task is to pass the XXVIII Amendment."

"Good lord, Jodi. You ever slow down to a gallop?"

Jodi sat back gingerly in the huge swivel chair.

"A President's chores are never done. Not much different than farm work, something city people rarely understand."

"Farming?"

"After you finish milking cows they're bawling loudly for you to do the job over again that evening. Farm animals are like humans. Once you pull away their independence and self-reliance you've got to take care of them every single day."

"Animals can't help themselves. They get captured."

"So do humans."

"By what?"

"Inertia."

Jodi played with the pen on her desk as Tony waited for her to continue. He knew there was more to come.

"Whether Americans pick up the ball on this amendment and run with it is now up to them. History shows there's been no guarantee that a society will endure once it becomes lazy and self-indulgent. From here on it's a tossup. We return to the dreams of our forefathers or succumb to the endless demands of ethnic drones. Over the centuries even the bees learned they can't survive with a hive of drones. They're expelled once they serve their purpose."

"You paint a dreary picture," said Tony.

"Sitting in a leaking boat and watching it sink is a coward's choice. People got to get off their duffs and make it seaworthy again."

"Perhaps they will once they wake up."

"They face the loss of their right to rule," said Jodi. "I face only a job loss from which I can return to Las Vegas and live like a Queen."

"Maybe Sir Alex Tytler was right 200 years ago."

"His belief that 'a democracy cannot exist as a permanent form of government' because it 'always collapses over loose fiscal policy' is right on target today," agreed Jodi. "And we're on the razor's edge of making it much past the 200 years he predicted."

Jodi sat eyeing Tony with a half-smile for several minutes. Tony squirmed uncomfortably in his seat.

"Something wrong, Madame President?" He only used that form of address when vexed with the games Jodi played with him.

"Wondering about you?"

"Wondering...?" repeated Tony, perplexed.

"Yes."

"What about me?" said Tony, defensively.

"Wondering why you're dense as a log?"

"Dense...why..." Tony sputtered, angry and hurt, blind to the wide smile on Jodi's face. "What have I done?"

"It's what you've not done?"

"Now what the hell...?"

"When are you going to propose? I'll never marry a man who doesn't first propose."

Tony's face went through more gyrations than a strangling turkey. When he finally caught his breath he managed.

"But you're the President?"

"So I've got to prod all the inert? Even my lover?"

"You know I love you."

"That's a given."

"I'd love to marry you in a minute. I've loved you for years."

"Doesn't sound like a proposal."

Tony scrambled to his feet and rushed to Jodi's side, dropping quickly on one knee.

"My God, Jodi. Will you marry me? Will you marry me? Will you marry me?"

Jodi laughed, leaning forward to kiss Tony on the forehead.

"Once is enough, Tony. I accept."

"Didn't want you to think I sought marriage because of your station," said a still confused Tony.

"I'm marrying character. Far more important than station," said Jodi.

57. THE CEREMONY

Jodi wanted to hold a private wedding in the small offices where she and Tony had first met. A later reception would take place at the White House. That turned out to be impractical for a President of the United States. Determined not to conduct a White House circus, she decided the wedding would take place in Las Vegas with Governor Aker giving away the bride. Since both her parents were dead, she invited Congressman Pace from Nebraska who had gone to bat for her first job on the hill.

When Senator Trudeau heard the news from Jodi, she hooted with delight.

"How wonderful...And I've news for you. General Hardin and I are to be married this month also. "Bout time. Been single over three years."

"Sounds like a Nevada uprising. Plan on getting hooked back home?" asked Jodi.

"Sure thing. The people from home elected me, not those within the beltway."

Jodi had a sudden inspiration.

"Want to go down in the history books, May?"

"Like getting shot twice? No thanks," said May suspiciously.

"A double wedding. The President and Vice President?"

Senator Trudeau nearly dropped the phone. A long period of silence followed.

"Great idea," she finally enthused, recovering her surprise and adding with a wry laugh. "Long as the President and Vice President aren't marrying each other."

"Our paramours have different tastes, thank God," laughed Jodi.

President Ericson let Tony and Senator Trudeau work out the wedding plans while she turned her attention to the amendment process.

"The XXVIII Amendment to the constitution is to be my legacy to the American People," Jodi said to Tony. "I must get it done. Otherwise we'll have revolution or dictatorship within the next two decades."

"Aren't you a bit fatalistic?"

"Nothing's more terrifying to the human spirit than to think you're in a situation from which you can't possibly escape," Jodi said. "For a cornered animal the choice is charge, risking death and possible freedom or cowering in a corner waiting certain death."

"I don't follow?"

Jodi spoke somberly.

"When government taxes people into a hopeless corner their only response is to choose a protector—by revolution or dictatorship."

Said Tony. "Invariably they end up worse. They get a Stalin, Mussolini or Hitler."

"A future carrot's always more promising than present misery," said Jodi. "A condition we're catapulting into today."

"I'll admit our elected royal body is incapable of restraint," said Tony.

Jodi shook her head sadly.

"A monarch's high taxes for frivolous projects set the stage for the bloody French Revolution; King George III precipitated ours when his demands were not one-tenth those enacted by congress; Hitler's rise followed an inflation and taxation binge; Russia's revolt against the Czars was an uprising of abused desperate humans driven into a corner."

Tony studied Jodi's flushed face. From the beginning he had known her war against congress was no shallow thing.

"There certainly must be some hope?"

"Yes. There's a window of hope. Initiated by Newt Gingrich into the Congress of '95. I'm trying to fulfill and expand that effort--so gutted and keelhauled as to become almost unrecognizable. He believed government money belongs to the people. And the people should better say how it's spent.

"Congress as its constituted never seems to learn," said Jodi. "They're not annointed gods to rule their fellow men. Nor are non-elected bureaucrats. They're to provide limited services which require cooperation, not dictatorial omniscience."

Quickly, she regained composure as the pain in her shoulder flashed a warning.

"Soon as I'm able I've got to hit the hustings again," she said with a warming smile. "You'll not be short-changed on the wedding bed. My need equals yours."

Tony doubted that but he said nothing.

*　　　*　　　*

The day of the wedding the Amendment had passed in 17 states with governors' signatures waiting in another 18. The few projected holdouts were not enough to spell defeat.

President Jodi Ericson viewed the crowd in high spirits as she exited from the church with Tony, May and Pete. The ceremony seemed interminable but had gone smoothly. Now she was President Jodi Pellini.

Her face broke into a bright smile as she reached up to whisper in Tony's ear.

"My ancestors would spin in their graves if they knew a Mongol had surrendered to an Italian."

58. EPILOGUE

The XXVIII Amendment and its twelve articles as eventually passed by the required number of three-fourth's of the states, read as follows:

1. Only one project at a time will be included in a bill for the President's signature or veto. Such bill shall not exceed 10 pages. When not possible, the President, at his discretion, may invoke the line item veto before signing a larger bill into law. No bill over 10 pages shall be submitted to members less than a full working week before vote.

2. A two-thirds vote of the senate shall be required to reject any Presidential appointment or any treaty negotiated and signed by the President of the United States.

3. Total federal government spending shall not exceed twenty percent of the gross national product (GNP); exception--Social Security, administered as an independent self-sufficient organization. Nor shall Congress raise taxes unless by a two-

thirds vote of both House and Senate and signed by the President. Exceptions to either spending or taxing totals are permitted only following a declaration of war by Congress or declaration of National Emergency by the President.

4. Money raised for one account shall not be diverted, borrowed or loaned to any other account, i.e., money raised for special purposes--such as airports, roads, immigration control--will not be moved into general operations or any other account. Excesses or shortages must be dealt with by separate legislation.

5. Universal conscription, involving all male youths eighteen through twenty years of age and such female youths as congress might later prescribe, shall take effect July 4 of the year following ratification. Term of service shall be for three years upon graduation from high school or starting from the eighteenth birthday. Only disabling medical exemptions will be permitted plus other personal restrictions, such as felons, as set forth by majority vote of both House and Senate and signed by the President. Furthermore, since all male youths shall be required to serve their country after graduating from high school, all voters must be capable of reading and writing English at that same high school level--the level required of the nation's soldiers.

6. Following ratification Congress shall reduce real and actual (not projected) spending--except as necessary in defense--by twenty percent each term to achieve a balanced budget. To do this no new taxes are authorized. Congressional salaries shall be reduced twenty percent each term the spending reduction is not met until total federal spending falls to twenty percent of GNP. Defense should stay constant at five percent of the twenty percent.

7. Terms of office for both elected and appointed officials shall be as follows: (a) Representatives shall be elected for three years. (b) Senators shall be elected for six years, with one-half standing for each election with members of the House.

Any interim appointments shall be for the term only of the replaced member. No representative or senator shall serve more than eighteen years. (c) The President's term shall be six years, so he will stand for election with all representatives and one-half the senators. (d) Supreme Court Justices and all Federal Judges shall be appointed for eighteen-year terms. No Justice shall serve beyond the age of eighty. (e) A two-thirds vote of the Senate, signed by the President, may remove any federal judge without cause or prejudice.

8. All U.S. Government agencies, organizations and quasi-government organizations will abide by due process, and are enjoined from placing requirements on state and local governments which require mandatory expenditures by states, local governments, jurisdictions or the people without complete funding or their consent.

Burden of proof for alleged violations shall lie with the accusing agency. Miranda-type warnings will be given in all civil as well as criminal contacts by the IRS, the EPA, or any agency of government.

9. Gerrymandering shall be prohibited. Congressional election districts shall be constructed for geographical compactness and ease of access. The Supreme Court is to oversee alignments.

10. No federal agency, except those mandated by the constitution, shall live beyond eighteen years without being recreated. A new law passed by Congress and signed by the President shall be required for further funding--after a thorough re-analysis of that agency's value to the nation.

11. Any state legislature, or citizens with a petition bearing five percent of the registered voters who voted in the last Presidential election, may put on the ballot a proposed constitutional amendment. If passed by a majority vote in three-

fourths of the states within six years, the proposed amendment shall become law.

12. The matter of pay: The federal bureaucracy shall be restructured. Explicit jobs will pay explicit wages, which scales shall be reviewed every six years after each Presidential election. Then increases or decreases shall be by majority vote of both houses and signed by the President. Interim increases in pay shall be the result of promotions or transfers to higher positions, not automatic increases. Those above grade GS-14 are considered political appointees subject to removal without cause or prejudice. Military and federal retirees shall not receive total federal retirement benefits higher than their active service pay in any single agency. Maximum pay in all U.S. Government agencies, organizations and quasi-government organizations shall not exceed triple the average pay of all U.S. workers.

Members of the House and Senate shall not receive one penny over the salaries paid them by the people. Exceptions will be in cases involving their own money, which assets they must have owned before taking oath of office. Nor are members to receive money from individuals or companies which directly or indirectly do business with any government agency--federal, state or local.

Campaign funds will be kept in separate audited accounts.

Jodi was convinced such passage had taken the first steps to put U.S. democracy on hold for another 200 years. Change was as necessary to the survival of government as to "the survival of the species." But it had to be the right kind of change. A blind change with modern atomic bombs would be catastrophic. An evolutionary change as achieved in our past would provide for hope of future "*survival of the species*." But now man's future survival has moved from physical to mental adaptation. Is he up to the challenge? Survival of the Species was never guaranteed.

ISBN 0-9623535-3-1

51795 >

EAN

9 780962 353536